MODEL THEORY OF
STOCHASTIC PROCESSES

Lecture Notes in Logic

A Publication of
The Association for Symbolic Logic

LECTURE NOTES IN LOGIC   14

MODEL THEORY OF STOCHASTIC PROCESSES

by

Sergio Fajardo
Department of Mathematics
University of Los Andes, Bogotá, Colombia

H. Jerome Keisler
Department of Mathematics
University of Wisconsin, Madison, Wisconsin, USA

ASSOCIATION FOR SYMBOLIC LOGIC

CRC Press
Taylor & Francis Group
Boca Raton London New York

CRC Press is an imprint of the
Taylor & Francis Group, an **informa** business

AN A K PETERS BOOK

CRC Press
Taylor & Francis Group
6000 Broken Sound Parkway NW, Suite 300
Boca Raton, FL 33487-2742

First issued in paperback 2019

ISBN 13: 978-1-56881-172-7 (pbk)

Visit the Taylor & Francis Web site at
http://www.taylorandfrancis.com

and the CRC Press Web site at
http://www.crcpress.com

Library of Congress Cataloging-in-Publication Data
Fajardo, Sergio, 1956-
    Model theory of stochastic processes / by Sergio Fajardo, H. Jerome Keisler.
      p. cm. – (Lecture notes in logic : 14)
    Includes bibliographical references and index.
    ISBN 1-56881-167-5 (acid-free paper) – ISBN 1-56881-172-1 (pbk. : acid-free paper)
      1. Stochastic processes. 2. Model theory. I. Keisler, H. Jerome. II. Title. III. Series.

QA274 .F37 2002
519.2'3–dc21   2002066210

Publisher's note: This book was typeset in LATEX, by the ASL Typesetting Office, from electronic files produced by the authors. using the ASL documentclass `asl.cls`. The fonts are Monotype Times Roman. The cover design is by Richard Hannus, Hannus Design Associates, Boston, Massachusetts.

Addresses of the Editors of Lecture Notes in Logic and a Statement of Editorial Policy may be found at the back of this book.

Association for Symbolic Logic:
C. Ward Henson, Publisher Mathematics Department,University of Illinois,1409 West Green Street, Urbana, Illinois 61801, USA

DEDICATION

Fajardo

Mi trabajo está dedicado a María Clara, Alejandro y Mariana, quienes me han acompañado y apoyado en la realización de mis sueños, incluyendo, por supuesto, este libro.

It may look strange, but I also want to dedicate this book to Jerry Keisler, just as a public excuse to say thanks for all I learned from him.

Keisler

Dedicated to Lois, my wife and closest friend. Thank you, Lois, for your love, for raising a wonderful family, and for your patience and understanding as my bridge between mathematics and the real world.

CONTENTS

INTRODUCTION

This book studies stochastic processes using ideas from model theory. Some key tools come from nonstandard analysis. It is written for readers from each of these three areas. We begin by intuitively describing this work from each of the three viewpoints.

From the viewpoint of probability theory, this is a general study of stochastic processes on adapted spaces based on the notion of adapted distribution. This notion is the analog for adapted spaces of the finite dimensional distribution, and was introduced by Hoover and Keisler [1984]. It gives us a way of comparing stochastic processes even if they are defined on different adapted spaces. A central theme will be the question

When are two stochastic processes alike?

There are several possible answers depending on the problem at hand, but our favorite answer is: Two stochastic processes are alike if they have the same adapted distribution. Early on in this book, we will consider questions of the following kind about an adapted space Ω, with the above meaning of the word "alike".

(1) *Given a stochastic process x on some other adapted space, will there always be a process like x on Ω?*

(2) *If a problem with processes on Ω as parameters has a weak solution, will it have a solution on Ω with respect to the original parameters?*

(3) *If two processes x, y on Ω are alike, is there is an automorphism of Ω which preserves measures and filtrations and sends x to y?*

Questions (1) – (3) ask whether an adapted space Ω is rich enough for some purpose. Adapted spaces with these properties are said to be universal, saturated, and homogeneous, respectively. Several arguments in probability theory can be simplified by working with a saturated adapted space, especially existence theorems which ordinarily require a change in the adapted space. In practice, probability theory allows great freedom in the choice of the adapted space. One does not care much which space is being used, as long as it is rich enough to contain the processes of interest.

Why not make life easier and work within a saturated adapted space when the situation calls for it? So far, this has rarely been done in the literature. The main reason is that none of the commonly used adapted spaces, such as the space of continuous paths with the Wiener measure and the natural filtration, are saturated. One of our objectives in this book is to correct this situation by making saturated adapted spaces more familiar to the reader, explaining where they come from and how they behave.

Throughout the book, we will proceed in parallel along two levels—the easier level of probability spaces with the language of finite dimensional distributions, and the harder level of adapted spaces with the language of adapted distributions. Most definitions and theorems at the probability level have analogs at the adapted level. As one would expect, new difficulties and ideas appear at the adapted level, where the filtration enters the picture.

From the point of view of model theory, we will work in a setting which is radically different from the classical setting, but the questions we ask will have a model theoretic flavor. As is often the case when ideas from model theory are applied in a new direction, the first step is to define a language that is adequate for expressing the main properties of the structures to be investigated. In this case the language is adapted probability logic and the structures are stochastic processes on adapted spaces. After the introduction of the language one proceeds to study the model-theoretic concepts suggested by the new language, and then explore their consequences for probability theory.

The main players in classical model theory are formulas of first order logic, algebraic structures, and elements of algebraic structures. Here, adapted functions take the place of formulas, adapted probability spaces take the place of algebraic structures, and stochastic processes (or random variables) take the place of elements of algebraic structures. The adapted distribution of a stochastic process is analogous to the elementary type of an element. The notions of universal, homogeneous, and saturated adapted spaces are closely analogous to, and were named after, the notions of universal, homogeneous, and saturated structures which are of central importance in model theory.

From the viewpoint of nonstandard analysis, our aim is to understand why there is a collection of results about stochastic processes which can only be proved by means of nonstandard analysis. Our explanation is that these applications require adapted spaces with properties such as saturation and homogeneity which are easy to get using nonstandard methods but difficult or impossible without using them.

As everyone who works in the subject knows, nonstandard analysis is not really a branch of mathematics in its own right, but is rather a general tool that can potentially be used in almost any field of mathematics. It is used in the same manner as, say, the language of topology, or the arithmetic of cardinals and ordinals. The method provides a natural approach for a great variety of problems. However, the proportion of mathematicians who are comfortable enough with the method to use it effectively in research is, and always has been, infinitesimal. For this reason, it has so far been used extensively in only a few areas, such as probability theory, Banach spaces, partial differential equations, and parts of mathematical physics and economics, with a few notable applications in other areas.

We now briefly describe the origin of the ideas in this book. The first paper to be mentioned is "Hyperfinite Model Theory" Keisler [1977], written in the middle 70's. The theme of this paper was to use hyperfinite models and nonstandard probability theory to handle large finite models in a systematic way. Thus contact was made between probability theory, nonstandard analysis, and model theory.

This period saw the beginnings of a long and successful series of applications of nonstandard probability theory to stochastic analysis. Loeb [1975] and Anderson [1976] wrote two important papers, showing respectively how to build standard probability spaces out of nonstandard ones, and a nonstandard (hyperfinite) construction of Brownian motion that captured the basic intuition that a Brownian motion is a random walk with infinitesimal steps. Keisler began work on applications of these new constructions, leading to the memoir "An Infinitesimal Approach to Stochastic Analysis", Keisler [1984], which was circulated in a preliminary version in 1979 and published in 1984. This memoir introduced a new method for proving existence of strong solutions of stochastic differential equations, and opened the door to further developments.

Adapted probability logic was the focus of a logic seminar at Wisconsin in 1979. The unpublished notes from that seminar, "Hyperfinite Probability Theory and Probability Logic" Keisler [1979], were circulated outside. The notion of adapted equivalence first appeared in these notes. The 1984 paper "Adapted Probability Distributions" by Douglas Hoover (a 1979 Ph.D. student of Keisler) and Keisler formulated the model-theoretic ideas of Keisler [1984] in purely probabilistic terms, and applied them to stochastic analysis. The subsequent work on the model theory of stochastic processes in the papers Hoover [1984] – Hoover [1991] greatly advanced the subject and is one of the main topics of this book.

We began writing this book during the fall of 1990 when Fajardo (a 1984 Ph.D. student of Keisler) was visiting Wisconsin and taught an advanced course on the subject. Preliminary notes were handed out, but new ideas appeared along the way. The series of papers starting with Keisler [1991] and continuing with Fajardo and Keisler [1996a] – Fajardo and Keisler [1995] and Keisler [1995] – Keisler [1997a] developed the theory of neometric spaces, with the goal of explaining the power of the methods and ideas from nonstandard analysis in standards terms. The results in Keisler [1995] and Keisler [1998] tied the new theory in with the previous work on the model theory of stochastic processes. This situation created a problem: we could not write the book on adapted probability logic as originally planned, but were forced to include new material to give the complete picture. This we did.

Prerequisites

To decide what to include in a book like this is a real headache, since we build upon three different areas.

With respect to probability theory. the reader does not have to be an expert, but some basic background is necessary. Adequate references are Ash [1972], Billingsley [1979], Williams [1991] and Ethier and Kurtz [1986].

We also assume some basic nonstandard probability theory. For this we rely on some very accessible works that very quickly prepare a reader for such an endeavor. Good references are Albeverio, Fenstad. Hoegh-Krohn, and Lindstrøm [1986], Stroyan and Bayod [1986], Lindstrøm [1988]. Keisler [1988], Fajardo [1990b] and Cutland [1983].

The following remark may come as a surprise: logic and model theory are not needed. We have stated the ideas from logic and model theory in a way that avoids all logic formalism. Of course, this was done on purpose. The point is that we are looking at the theory of stochastic processes from the point of view of a model theorist. which is a different approach but need not be considered a work in logic. The book is self contained with respect to ideas of a model theoretic nature. All the new and relevant concepts are presented.

A reader going through the book can ignore some details and still get a fairly good view of the subject. We believe this book will be of interest to mathematicians willing to explore new developments with an open mind.

Acknowledgements

Fajardo: During the period of work on this book I have received and enjoyed the support and help of several institutions and friends. In Colombia I have always had full support from the School of Sciences at the Universidad de los Andes. They gave me the time I needed. and in a few occasions helped with financial support. But the most important contribution has been the quality of the students I have had. A few of them have written their undergraduate and masters thesis under my direction. working on problems and ideas directly related to the topics treated in the book. Most of them have gotten their Ph.D degrees in Mathematics in the United States. Here are their names: Miguel López. Carlos Andrés Sin. Javier Peña, Emilio Remolina and Ramiro De la Vega.

Two other Institutions gave me support in order to carry out this work: El Banco de la República and Colciencias. I am indebted to them for believing in this type of project, that is unusual in the Colombian context.

Keisler: I wish to thank the many friends and colleagues who have provided a stimulating mathematical environment and valuable insights which made this work possible. Among these colleagues are Bob Anderson, Nigel Cutland. Ward Henson, Doug Hoover, Tom Kurtz. Peter Loeb, and David Ross. I gratefully acknowledge the generous support I received from the National Science Foundation and the Vilas Trust Fund, which allowed us to get together several times to work on this book.

CHAPTER 1

ADAPTED DISTRIBUTIONS

One of the very first questions addressed by researchers within each particular area of mathematics is: When are different objects in the theory "alike"? Answers, of course, depend on the nature and specific conditions of the subject under study. Let us begin this book by taking a brief look at the way this question is dealt with in probability theory.

At the foundational level, the basic objects studied in probability theory are probability spaces and random variables. Usually, random variables are defined on totally unrelated sample spaces, so there is no point in comparing them as pointwise functions. When studying a random variable, what matters is its probability distribution, or distribution law. Intuitively, the distribution law of a random variable gives us the probabilities of the occurrence of events related to the variable. This point of view leads to a simple and natural notion of sameness, which we will review in the next section.

1A. Probability distributions

Two random variables, regardless of the probability space where they are defined, are "alike" if they have the same distribution law (or more briefly, the same law). Since the law is an important concept that underlies some of the main ideas of our work, this is a good place to review it and fix some notation.

DEFINITION 1A.1. *Let* $(\Omega, \mathcal{F}, P)$ *and* $(\Gamma, \mathcal{G}, Q)$ *be probability spaces, and let* x *and* y *be real valued functions defined on* Ω *and* Γ *respectively.*

(*a*) x *is a* **random variable** *if for every Borel subset* A *of* $\mathbb{R}$, *the set* $x^{-1}(A) \in \mathcal{F}$.

(*b*) *The* **distribution**, *or* **law**, *of* x *is the probability law* (x) *defined on the Borel* σ*-algebra* $\mathcal{B}(\mathbb{R})$ *by the formula*

$$law\,(x)[A] = P[x \in A] = P[x^{-1}(A)].$$

(*c*) *Two random variables* x *and* y *are said to* **have the same distribution (or law)** *if* $law\,(x) = law\,(y)$. *We denote this relation by* $x \equiv_0 y$.

Random variables do not have to be real valued. We also allow them to take values in a Polish space M. (In this book, **Polish space** always means complete separable metric space). The only difference in the above definition is that instead

1

of taking a Borel subset of $\mathbb{R}$ we take a Borel subset of M. In general, we use the phrase "random variable" without bothering to mention the space where the function takes its values, unless we need to make the target space explicit.

In the probability literature one often reads about random variables which are normally distributed, have the Poisson distribution, etc., without mentioning the space where they live, since the useful and interesting results are derived from the properties of the distributions.

The next step in probability theory is the notion of a stochastic process which, very roughly, is a random variable that models the notion of random evolution through time. Again, it is known from basic probability theory that what really matters when studying stochastic processes is the family of finite dimensional distributions attached to each process. This notion extends the notion of a probability distribution of a random variable in a natural way. Let us now introduce some of the main concepts needed for the study of stochastic processes in this book.

For simplicity, we will usually restrict our attention to stochastic processes with the time index set $[0, 1]$, with the understanding that $[0, 1]$ can be replaced with minor changes by sets like $[0, \infty)$ or $[0, \infty]$.

DEFINITION 1A.2. (a) *A real valued* **stochastic process** *x defined on a probability space $(\Omega, \mathcal{F}, P)$ is a function $x : \Omega \times [0, 1] \to \mathbb{R}$ such that for each $t \in [0, 1]$, the function $x(\cdot, t) : \Omega \to \mathbb{R}$ is an $\mathcal{F}$-measurable random variable. Sometimes we use the notation $x = (x_t)_{t \in [0,1]}$, or $x = (x(t))_{t \in [0,1]}$. The set $[0, 1]$ is called the* **time-index set**.

(b) *x is said to be a* **measurable** *stochastic process if as a function on the product space $\Omega \times [0, 1]$ it is measurable with respect to the product σ-algebra $\mathcal{F} \times \mathcal{B}$, where $\mathcal{B}$ is the Borel σ-algebra on $[0, 1]$. Unless otherwise stated, all processes in this book are assumed to be measurable.*

(c) *Let x be a real valued process defined on a probability space $(\Omega, \mathcal{F}, P)$. The* **finite dimensional distribution** *of x is the unique family of probability measures $P^x_{t_1, \ldots, t_n}$ on the σ-algebras $\mathcal{B}(\mathbb{R}^n)$ such that for all times $t_1, \ldots, t_n$ and Borel sets $A_1, \ldots, A_n$ in $\mathcal{B}(\mathbb{R})$, we have*

$$P^x_{t_1 \ldots t_n}[A_1, \ldots, A_n] = P[x_{t_1} \in A_1, \ldots, x_{t_n} \in A_n].$$

(d) *Let $(\Omega, \mathcal{F}, P)$ and $(\Gamma, \mathcal{G}, Q)$ be probability spaces and x and y be real valued processes defined on Ω and Γ respectively. x and y have the same finite dimensional distribution, in symbols $x \equiv_0 y$, if*

$$P^x_{t_1 \ldots t_n} = Q^y_{t_1 \ldots t_n} \text{ for all times } t_1, \ldots, t_n.$$

There is one more way we can think of a stochastic process which is useful in some situations. We can look at it as a random variable $x : \Omega \to M$, where M is the set of functions from $[0, 1]$ into $\mathbb{R}$. For a given ω, the function $x(\omega) : [0, 1] \to \mathbb{R}$ defined by $x(\omega)(t) = x(\omega, t)$ is called the **path**, or **trajectory**, of x at ω. Depending on what properties the process has or what type of process we want to study we can put a topology on M so that it becomes a Polish space.

and then x is just a random variable on Ω. A familiar example is the case of a continuous stochastic process, which is the same thing as a random variable on Ω with values in the space $C([0, 1], \mathbb{R})$ with the metric of uniform convergence.

As with random variables, one reads in the literature about stochastic processes with specific properties (for example, Poisson processes, Markov processes, Brownian motions), without even mentioning the space where these processes are supposed to be defined. These properties typically depend only on the finite dimensional distribution of the process. There is a famous theorem in the literature known as Kolmogorov's Existence Theorem (see Billingsley [1979] or Lamperti [1977]) that is worth mentioning here.

Given a stochastic process x, it is easy to verify that the finite dimensional distribution

$$\mathcal{P}^x = \{P^x_{t_1 \ldots t_n} : n \in \mathbb{N}, t_i \in [0, 1]\}$$

is a family of probability measures on the $\mathcal{B}(\mathbb{R}^n)$'s which satisfies the following consistency conditions:

Consistency Conditions

(a) If π is a permutation of $\{1, 2, \ldots, n\}$ and ϕ_π is the function from $\mathbb{R}^n \to \mathbb{R}^n$ defined by

$$\phi_\pi(x_1, \ldots, x_n) = (x_{\pi(1)}, \ldots, x_{\pi(n)})$$

then

$$P^x_{t_{\pi(1)} \ldots t_{\pi(n)}}[C] = P^x_{t_1 \ldots t_n}[\phi_\pi^{-1}(C)].$$

(b) If $\phi_{n+m,n}$ is the projection function from $\mathbb{R}^{n+m}$ onto $\mathbb{R}^n$ then

$$P^x_{t_1 \ldots t_n}[C] = P^x_{t_1 \ldots t_{n+m}}[\phi_{n+m,n}^{-1}(C)].$$

Theorem 1A.3. (**Kolmogorov Existence Theorem**) *Given a family*

$$\mathcal{Q} = \{\mu_{t_1 \ldots t_n} : n \in \mathbb{N}, t_i \in [0, 1]\}$$

of probability measures on the corresponding $\mathcal{B}(\mathbb{R}^n)$'s that satisfies the consistency conditions above, there exists a probability space with a stochastic process y whose finite dimensional distribution $\mathcal{P}^y$ is $\mathcal{Q}$. ⊣

Kolmogorov's Existence Theorem is important because it allows one to show that certain stochastic processes exist, just by giving some conditions that must be satisfied by their finite dimensional distributions. There are important results in the probability literature which are direct consequences of this theorem, such as the existence of a Brownian motion (see Billingsley [1979]) or stationary Markov chains (see Lamperti [1977]).

1B. Adapted spaces

Probability theory has continued to expand its scope. The last three decades have seen progress in what is sometimes called the General Theory of Stochastic Processes. This theory was pioneered by Doob, Levy and Kolmogorov, among others. See, for example, the books Dellacherie and Meyer [1978], Dellacherie and Meyer [1988], or Ethier and Kurtz [1986]. The intuitive idea is that the structures of probability spaces can be enriched in order to capture the notion of information that accumulates over time. This intuition is formalized by adding to the structure of a basic probability space a family of σ-algebras, one for each time t, whose events are the information that has been gathered up to that time. In this way, stochastic processes are related to that information via the conditional expectation operation, and many more phenomena can be modelled. Here are the definitions.

DEFINITION 1B.1. (a) *An* **adapted space** Ω *is a structure of the form*

$$\Omega = (\Omega, \mathcal{F}_t, P).$$

where $(\Omega, \mathcal{F}_1, P)$ *is a probability space and* $(\mathcal{F}_t)_{t \in [0,1]}$ *is an increasing family of* σ-*algebras, called a* **filtration**, *satisfying the following conditions* (*called the* **usual conditions**):

Right continuity. For each $t < 1$, $\mathcal{F}_t = \bigcap_{s > t} \mathcal{F}_s$.
Completeness. $\mathcal{F}_0$ *is* P-*complete.*

(b) *We say that a stochastic process* x *on* Ω *is* **adapted to** $(\mathcal{F}_t)$ *if for each* t, x_t *is* $\mathcal{F}_t$- *measurable.*

Once we are dealing with stochastic processes which live on adapted spaces, the following natural question comes up. Is there a notion for adapted spaces that is analogous to random variables having the same distribution, or to stochastic processes having the same finite dimensional distribution? Or, on a more intuitive level, can we capture the idea of two stochastic processes on adapted spaces describing the same phenomenon, or having the same probabilistic behavior? The unpublished article "Weak convergence and the general theory of processes", Aldous [1981], presents in its introduction an interesting probabilistic argument calling for a notion like the one we are going to introduce in this chapter. Here is an excerpt of his point of view:

"... there is a more fundamental question to be resolved: when we regard processes in the Strasbourg manner (i.e., living on adapted spaces), what should we mean by their 'distributions'? In other words, what characteristic, analogous to distribution in the function space theory, contains the essential description of the process and can be used to tell whether two processes are essentially the same or not?"

In this chapter we propose an answer to these questions. We introduce the notion of adapted distribution which appeared in Keisler [1979] and Hoover and Keisler [1984], and develop the basic features of this new concept. It is important to point out that some probabilists, including Aldous [1981], have suggested their own answers to these same questions. Later on we will see how they fit within our current framework; see for example Jacod [1979] and the comments in the introduction of Hoover [1992].

As we explained in the Introduction, the underlying idea of the project we are about to undertake is as follows. Define a language that is adequate for expressing the properties of stochastic processes on adapted spaces. After the introduction of the language, define notions such as elementary equivalence, proceed to build the model theory associated with the new language, and then explore its consequences in probability theory.

We will introduce the language in a way that avoids the logical formalism. Our intent is to show how the probability concepts suggest the model theoretic treatment. A more traditional logical presentation, where the reader can find axiom sets and the completeness theorems that have been proved for the host of probability logics that have appeared since the publication of the paper Keisler [1977], can be found in Keisler [1985], Fajardo [1985a] and the monograph Raškovic and Dordevic [1996] of Raškovic and Dordevic.

In model theory, the language is a vehicle that allows us to compare mathematical objects, possibly defined over very different universes, that share a basic structure. Moving from random variables to stochastic processes, it was natural to extend the notion of having the same distribution to the notion of having the same finite dimensional distribution. What can we do now that filtrations have to be taken into account?

Well-known characterizations of the concepts of equality in distribution and same finite dimensional distribution give us a good hint.

PROPOSITION 1B.2. (*a*) *Two random variables x and y have the same distribution if and only if* $E[\phi(x)] = E[\phi(y)]$ *for every bounded continuous function* $\phi : \mathbb{R} \to \mathbb{R}$.

(*b*) *Two stochastic processes x and y have the same finite dimensional distribution if and only if*

$$E[\phi(x_{t_1}, \ldots, x_{t_n})] = E[\phi(y_{t_1}, \ldots, y_{t_n})]$$

for each n, all times $t_1, \ldots, t_n \in [0, 1]$, *and each bounded continuous function* $\phi : \mathbb{R}^n \to \mathbb{R}$.

(*c*) *A sequence* (x_n) *of random variables weakly converges to a random variable x if and only if* $E[\phi(x_n)]$ *converges to* $E[\phi(x)]$ *for every bounded continuous function* $\phi : \mathbb{R} \to \mathbb{R}$.

The content of this proposition can be stated in a few words. Two stochastic processes x and y have the same finite dimensional distribution if after applying bounded continuous functions to the random variables x_t and y_t and then taking expected values, the same results are obtained. We are going to extend this idea to processes that live on adapted spaces. But before that we will make some needed remarks.

(1) As mentioned after Definition 1A.1, random variables do not necessarily have to be real valued, but may take values in Polish spaces M. Most of the time, but not always, the real valued case is enough for our purposes and the

general Polish case is just a straightforward extension. For example, the previous proposition can be restated for the Polish space case by making the following adjustment. Take bounded continuous functions ϕ from M^n into $\mathbb{R}$ instead of from $\mathbb{R}^n$ into $\mathbb{R}$. Then everything goes through.

Let us introduce some notation. The set of bounded continuous functions $\phi : M \to \mathbb{R}$ is denoted by $C_b(M, \mathbb{R})$. The set of M-valued random variables defined on a space Ω is denoted by $L^0(\Omega, M)$, following the usual convention of identifying random variables that are almost surely equal.

(2) In Definition 1 we introduced adapted spaces where time was indexed by the interval $[0, 1]$ and the filtration is right continuous (which, as indicated, is going to be the usual case in this book). As an alternative, one can take a step by step approach, increasing the complexity of the time index set as we go along, and defining adapted spaces in a way that will be used in Chapters 7 and 8. In all cases the filtration is required to be increasing and complete, but we do not require that it be right continuous except in the case that the time index set is a real interval. There are several natural time index sets which one might consider.

(2a) A finite time index set.

(2b) The set of natural numbers $\mathbb{N}$ (discrete time).

(2c) The set $\mathbb{B}$ of dyadic rationals in $[0, 1]$.

(2d) We can go as far as considering arbitrary linearly ordered sets $\langle \mathbb{L}, \leq \rangle$ with an initial element 0 as the time index set, and use the convention that $t < \infty$ for all $t \in \mathbb{L}$. Of course, all the cases presented so far are particular instances of this one.

(3) There are two ways in which Proposition 1B.2 (b) above can be viewed. The one suggested there is that we have a random variable

$$\phi(x_{t_1}, \ldots, x_{t_n}) : \Omega \to \mathbb{R}$$

for each tuple of times $(t_1, \ldots, t_n)$. Another alternative would be to think of times as variables and then get $\phi(x) : \Omega \times [0, 1]^n \to \mathbb{R}$ as an n-fold stochastic process, that is, an element of $L^0(\Omega \times [0, 1]^n, \mathbb{R})$. This difference, which is not essential for the moment, will lead to two different approaches to the theory of adapted distributions in this book.

1C. Adapted functions

Here is the intuitive idea of what has to be done now. In order to compare two stochastic processes x and y on adapted spaces one operates on the random variables x_t and y_t by means of bounded continuous functions as in the case of the finite dimensional distribution, and adds one extra feature. Take conditional expectations with respect to the σ-algebras from the filtration, in order to relate the process to the information that these σ-algebras carry. Then take expected values, and if the same results are obtained, the processes are said to have the same adapted distribution.

We are now ready for the main idea in what we have called the Model Theory of Stochastic Processes. The following notion of an adapted function is essentially in Hoover [1987]. It is closely related to the notion of a conditional process in Keisler [1979] and Hoover and Keisler [1984], which we will introduce a little later. As is commonly done in mathematical logic, we will introduced an iterated family of mathematical objects with an inductive definition.

DEFINITION 1C.1. *Let x be a stochastic process defined on an adapted space $\Omega = (\Omega, \mathcal{F}_t, P)$. The set AF of **adapted functions** is the family of expressions obtained by applying the following rules (i)–(iii) finitely many times. The value of an adapted function at (x, Ω) is a random variable in $L^0(\Omega, \mathbb{R})$ which is also given by these rules.*

(i) (Basis Step) If $\phi \in C_b(\mathbb{R}^n, \mathbb{R})$ and $\vec{t} = (t_1, \dots, t_n) \in [0, 1]^n$, then the expression $\phi_{\vec{t}}$ is in AF.

Its value at x is the random variable

$$\phi_{\vec{t}}(x) : \Omega \to \mathbb{R},$$

defined by

$$\phi_{\vec{t}}(x)(\omega) = \phi(x_{t_1}(\omega), \dots, x_{t_n}(\omega)).$$

(ii) (Composition Step) If $\psi \in C_b(\mathbb{R}^n, \mathbb{R})$ and $f_1, \dots, f_n$ are in AF, then the expression $\psi(f_1, \dots, f_n)$ is in AF. If the value of f_i at x is $f_i(x)$, then the value of $\psi(f_1, \dots, f_n)$ at x is the random variable

$$\psi(f_1, \dots, f_n)(x) : \Omega \to \mathbb{R},$$

defined by

$$\psi(f_1, \dots, f_n)(x)(\omega) = \psi(f_1(x)(\omega), \dots, f_n(x)(\omega)).$$

(iii) (Conditional Expectation Step) If f is in AF and $t \in [0, 1]$, then the expression $E[f|t]$ is in AF. Its value at x is the random variable

$$E[f|t](x) : \Omega \to \mathbb{R}$$

defined almost surely by

$$E[f|t](x)(\omega) = E[f(x)|\mathcal{F}_t](\omega).$$

Since only bounded continuous functions are used in the Basis Step in the definition of adapted functions, the conditional expectations are always finite and each adapted function has a uniform bound that does not depend on x. It follows by induction that for each adapted function f and random variable x on Ω, the value $f(x)$ is an almost surely well-defined random variable on Ω. It would make no difference if we allowed continuous functions instead of bounded continuous functions in the Composition Step (ii), because the image of a bounded set under a continuous function is automatically bounded. However, we need the functions in the Basis Step to be bounded to guarantee that the conditional expectations exist.

The definition of an adapted function in the paper Hoover [1987] was actually a bit simpler than the above definition, with only a single time instead of an n-tuple

of times in the Basis Step. This works because every adapted function in the
present sense can be approximated by adapted functions with only single times in
the Basis Step. We decided to use the slightly more complicated definition here
because it matches the criterion in Proposition 1B.2 for having the same finite
dimensional distribution, as well as the definition of a conditional process which
will come later.

Here is a typical example of an adapted function. Let $s, t \in [0, 1]$. Then the
expression

$$f = (E[\sin_s |t])^2$$

is an adapted function, and its value at a stochastic process x is the random
variable

$$f(x)(\omega) = (E[\sin(x_s)|\mathcal{F}_t])^2(\omega).$$

In this example the conditional expectation is taken before squaring. Squaring
before taking the conditional expectation would give a different adapted function
which has a different value at x. We can now proceed to take the expected value
of this random variable, and get a real number which depends on x, in this case

$$E[f(x)] = E[(E[\sin(x_s)|\mathcal{F}_t])^2].$$

Notice that the adapted function f by itself does not mention a stochastic
process x or space Ω at all. The value $f(x)$ is a random variable on the adapted
space Ω. Given any other stochastic process y on an adapted space Γ, the value
$f(y)$ is a random variable on Γ, and the expected value $E[f(y)]$ is again a real
number.

With this definition we can now compare stochastic processes defined on (possibly) different adapted spaces.

DEFINITION 1C.2. *Let x and y be stochastic processes on adapted spaces Ω and
Γ respectively. We say that x and y have the same **adapted distribution** if for each
f in AF we have $E[f(x)] = E[f(y)]$. which means $\int_\Omega f(x)dP = \int_\Gamma f(y)dQ$.
Denote this relation by $(\Omega, x) \equiv (\Gamma, y)$. or just by $x \equiv y$ if the underlying adapted
spaces Ω and Γ are understood. The relation $\equiv$ is called **adapted equivalence**.*

There are several possible reformulations and extensions of the notion of
adapted equivalence. Depending on the context any one of them can be used.

EXTENSIONS AND REFORMULATIONS.

(a) Sometimes we are interested in considering adapted functions built with a
time set $\mathbb{L}$ other than $[0, 1]$. In this case. we take times from $\mathbb{L}$ instead of $[0, 1]$
in Steps (i) and (iii) of Definition 1C.1. Adapted functions with times from $\mathbb{L}$
are called $\mathbb{L}$-**adapted functions**, and the set of $\mathbb{L}$-adapted functions is denoted by
$AF(\mathbb{L})$. We say that x and y have the same $\mathbb{L}$-adapted distribution if the above
definition holds for all f in $AF(\mathbb{L})$, and we write $x \equiv_\mathbb{L} y$.

(b) As we indicated in the preceding section, everything done so far can be
extended to processes taking values in a Polish space M rather than in $\mathbb{R}$. The
only necessary change is in the Basis Step. where we use $\phi \in C_b(M^n, \mathbb{R})$ instead

of $\phi \in C_b(\mathbb{R}^n, \mathbb{R})$. Hoover [1992] goes even further in relaxing the conditions imposed on M, but we do not need such generality in this book.

(c) Random variables with values in M can be regarded as particular stochastic processes with values in M which depend only on ω and not on time. Thus the notions of adapted function and adapted equivalence can be applied to random variables as well as to stochastic processes with values in M. In this case the times will not be needed in the Basis Step of the definition of adapted function, but times will be introduced in the Conditional Expectation Step. Observe that if x is a random variable and $x \equiv y$, or even $x \equiv_0 y$, then y is also a random variable.

(d) The definition of $x \equiv y$ can be extended to cases where more than one stochastic process is considered at a time. Let us consider the case of two processes. We want to define $(x, x') \equiv (y, y')$. We present two different ways in which this can be done when the processes take values in the Polish space M.

(d1) Taking into account the extension presented in (b), the problem is solved by simply considering the pairs (x, x') and (y, y') as M^2-valued stochastic processes, since M^2 is a Polish space. Following (b), adapted equivalence from this point of view is denoted as usual by $(x, x') \equiv (y, y')$.

(d2) Another approach is to modify the Basis Step. Given $\phi \in C_b(M^n, \mathbb{R})$ and $t \in [0, 1]^n$, introduce two adapted functions ϕ_t^1 and ϕ_t^2 whose corresponding values for the pair (x, x') are as follows. $\phi_t^1(x, x')$ is $\phi_t(x)$, and $\phi_t^2(x, x')$ is $\phi_t(x')$. The Basis Step will now say that:

If $\phi \in C_b(M^n, \mathbb{R})$ and $t \in [0, 1]^n$, then ϕ_t^1 and ϕ_t^2 are in AF.

The rest remains unchanged. In order to avoid confusion we temporarily denote this equivalence relation by $(x, x') \equiv' (y, y')$. This relation can be thought of as equivalence in a language which can handle two stochastic processes at a time on each adapted space.

There are some important points to explain here. The form we use for the two processes case can be extended to a finite number of processes in the obvious way. It can be easily proved (exercise) that both approaches are equivalent. One might wonder how far can we go with these extensions. If we have countably many processes at a time, there is no problem. But observe that in (d1) we would deal with $M^{\mathbb{N}}$, and the topology that we put on this space is important. The proof that both approaches are equivalent now requires more sophistication, since it uses the Stone-Weierstrass theorem and one has to understand how bounded continuous functions on $M^{\mathbb{N}}$ can be expressed as limits of functions that only depend on finitely many components.

What happens if we have uncountably many processes at a time? The alternative (d1) does not work since the target space would be of the form M^J with an uncountable J and may not even be metrizable. But (d2) still works since we always keep the same target space and just modify the Basis Step by adding one adapted function symbol ϕ_t^α for each ϕ, each tuple of times $t \in [0, 1]^n$, and each $\alpha \in J$.

(e) Hoover [1991] has yet another approach to adapted distribution, an extension of Knight's prediction process (see Knight [1975]). This approach turns out to be useful for studying convergence in adapted distribution. We will use a variant of this approach in Chapter 3. Section D, to give a characterization of adapted equivalence which uses only the notion of a Markov process and does not mention adapted functions.

As explained before, there are two different ways of handling adapted functions. treating time either as a free variable or as a constant. In Definition 1C.2. time is constant, so the values of the adapted functions on x are elements of $L^0(\Omega. \mathbb{R})$.

The other alternative. where time is treated as a free variable. changes the nature of the value of an adapted function at a process. Instead of a random variable it gives an n-fold stochastic processes where the n tells us how many time variables were used in the definition of the adapted function. This point of view was the one originally presented in the paper Hoover and Keisler [1984], and is needed in many applications of nonstandard analysis to continuous time stochastic processes. The expressions in the variable time approach which give n-fold stochastic processes are called n-fold conditional processes. Since this approach involves a slightly different way of looking at the notion of adapted equivalence, a careful definition is given here.

DEFINITION 1C.3. *Let x be a stochastic process on an adapted space Ω. The class CP of* **conditional processes** *is the family of expressions obtained by applying the following rules (i)–(iii) finitely many times. The value of an n-fold conditional process at x is an n-fold stochastic process which is also given by these rules.*

(i) (Basis Step) If $\phi \in C_b(\mathbb{R}^n, \mathbb{R})$, then f_ϕ is an n-fold conditional process. Its value at x is the n-fold stochastic process

$$f_\phi(x) : \Omega \times [0,1]^n \to \mathbb{R}$$

defined by

$$f_\phi(x)(\omega, t_1, \dots, t_n) = \phi(x_{t_1}(\omega), \dots, x_{t_n}(\omega)).$$

(ii) (Composition Step) If $\psi \in C_b(\mathbb{R}^k, \mathbb{R})$ and $f_1, \dots, f_k$ are n-fold conditional processes, then the expression $\psi(f_1, \dots, f_k)$ is an n-fold conditional process. If the value of each f_i at x is $f_i(x)$, the value of $\psi(f_1, \dots, f_k)$ at x is $\psi(f_1(x), \dots, f_k(x))$.

(iii) (Conditional Expectation Step) If f is an n-fold conditional process, then the expression $E[f \mid t]$ is an $(n+1)$-fold conditional process. If the value of f at x is the n-fold stochastic process $f(x)$, then the value of $E[f \mid t]$ at x is the $(n+1)$-fold stochastic process defined almost surely by

$$E[f \mid t](x)(t_1, \dots, t_n, t, \omega) = E[f(x)(t_1, \dots, t_n, \omega) \mid \mathcal{F}_t].$$

In order to make Step (ii) more readable, we did not write down the time and sample point variables $t_1, \dots, t_n, \omega$.

One can build two-fold stochastic processes either by composing one-fold stochastic processes with different time variables in Step (ii), or by taking a conditional expectation on a new time variable in Step (iii). For example, using the Composition Step we can build a conditional process f whose value at x is the two-fold stochastic process

$$f(x)(s, t, \omega) = \sin(x_s(\omega)) + \cos(x_t(\omega)).$$

And using the Conditional Expectation Step we get a conditional process g whose value at x is the two-fold stochastic process

$$g(x)(s, t, \omega) = E[\sin(x_s(\omega)) | \mathcal{F}_t].$$

In this approach, the equivalence $x \equiv y$ is defined by simply saying that for each n-fold f in CP and all times $(t_1, \dots, t_n)$ in $[0, 1]^n$ we have

$$E[f(x)(t_1, \dots, t_n)] = E[f(y)(t_1, \dots, t_n)].$$

We point out the key difference between the adapted functions and conditional processes. In the Basis Step, we have a different adapted function ϕ_t for each time constant $t \in [0, 1]$, and the value $\phi_t(x)$ at x is the random variable x_t. On the other hand, there is only one conditional process f_ϕ, and its value $f_\phi(x)$ at x is the stochastic process x_t where the symbol t is a variable.

From now on we will freely use both adapted functions and conditional processes, depending on which approach is best for the particular situation we have at hand.

Now it is time to put together several basic facts about the notions introduced so far. The proofs are good exercises which are left to the reader.

BASIC FACTS 1C.4. *Let x and y be stochastic processes on adapted spaces Ω and Γ.*

(a) $x \equiv y$ if and only if $x^m \equiv y^m$ for all $m \in \mathbb{N}$, where

$$x^m(\omega, t) = \min(x(\omega, t), m).$$

(b) $x \equiv_{AF} y$ if and only if $x \equiv_{CP} y$. Here we are just stating that the adapted function approach and the conditional process approach are equivalent.

(c) In the notation of part (d) in the list of Extensions and Reformulations, $(x, x') \equiv (y, y')$ if and only if $(x, x') \equiv' (y, y')$. The same is true with finitely many and countably many coordinates.

(d) Let $(x_n)_{n \in \mathbb{N}}$ and $(y_n)_{n \in \mathbb{N}}$ be two countable sequences of M-valued stochastic processes. Define $\bar{x}$ as the $M^{\mathbb{N}}$-valued process $\bar{x}(\omega, t) = (x_n(\omega, t))_{n \in \mathbb{N}}$, and similarly for $\bar{y}$. Then the following are equivalent.

(i) $\bar{x} \equiv \bar{y}$
(ii) $(x_1, x_2, \dots, x_n, \dots) \equiv' (y_1, y_2, \dots, y_n, \dots)$
(iii) $(x_1, x_2, \dots, x_n) \equiv' (y_1, y_2, \dots, y_n)$
for all $n \in \mathbb{N}$.

(e) Regardless of the approach, $x \equiv y$ if and only if $x \equiv_{\mathbb{L}} y$ for every countable time set $\mathbb{L} \subseteq [0, 1]$.

(f) If we have sequences $(x_n)_{n \in \mathbb{N}}$ and $(y_n)_{n \in \mathbb{N}}$ such that

$$(x_1, x_2, \dots, x_n, \dots) \equiv (y_1, y_2, \dots, y_n, \dots)$$

and there exists a random variable x such that $\lim x_n = x$ a.s., then there exists a random variable y such that $\lim y_n = y$ a.s. and

$$(x_1, x_2, \ldots, x_n, \ldots, x) \equiv (y_1, y_2, \ldots, y_n, \ldots, y).$$

(g) Let x_n, x be stochastic processes on Ω and y_n, y be stochastic processes on Γ such that $x_n \equiv y_n$ for each n, and for each $t \in [0, 1]$, $x_n(t) \to x(t)$ and $y_n(t) \to y(t)$ in probability. Then $x \equiv y$.

Part (g) is useful in convergence arguments, and is proved as Proposition 2.20 in Hoover and Keisler [1984]. It also holds with a.s. convergence instead of convergence in probability.

Before we begin to explore the new concepts, let us prove some results that tell us we are on the right track. For the remainder of this chapter it will always be understood that we are working on an adapted space Ω.

PROPOSITION 1C.5. *If $x \equiv y$ and x is a martingale then y is also a martingale.*

PROOF. We first have to verify that each y_t is integrable. Let us see how this is done. Let ϕ^n be the truncation of the absolute value function at n, so that ϕ_t^n is in AF. Observe that $\phi_t^n(x) \to |x_t|$ and the same is true with y in place of x. By hypothesis, for each n we have $E[\phi^n(x)] = E[\phi^n(y)]$. These two properties together with fact that x_t is integrable and the dominated convergence theorem give us that

$$\int |x_t| dP = \lim \int \phi_t^n(x) dP = \lim \int \phi_t^n(y) dQ = \int |y_t| dQ < \infty.$$

We now show that for each $s \leq t$, $E[y_t | \mathcal{G}_s] = y_s$. This property follows immediately from the corresponding property for x, which can be written

$$\int_\Omega |E[x_t | \mathcal{F}_s] - x_s| dP = 0.$$

To be rigorous, the left side of this equation is not captured in AF. But the equation can be proved using adequate truncations as in the preceding paragraph. Therefore one gets that

$$\int_\Gamma |E[y_t | \mathcal{G}_s] - y_s| dQ = 0,$$

which is exactly what we have to prove.   ⊣

This proposition shows something important: The property of being a martingale is preserved under $\equiv$. This is the least we could expect, since the notion of a martingale is the cornerstone of the general theory of processes.

At the same time, the proof tells us that the restriction to bounded continuous functions in the definition of AF usually can be sidestepped with a limit argument like the one above, and consequently all continuous functions are at our disposal. However, we cannot relax the boundedness restriction since we need to guarantee that all adapted functions are integrable.

Here is one more question:

PROBLEM 1C.6. *Is there an analog of the Kolmogorov Consistency Theorem 1A.3 for adapted distributions?*

1D. Adapted rank

A natural question comes to mind after introducing adapted equivalence. What is its relationship with the notion "x and y have the same finite dimensional distribution?" We shall see that adapted equivalence implies having the same finite dimensional distribution. But first we introduce another useful concept in this context, which in logic is known as the quantifier rank of a formula. Intuitively, we count the number of times the conditional expectation is used in the construction of an adapted function f.

DEFINITION 1D.1. (*a*) *For each adapted function f, the rank of f, denoted rank(f), is defined by induction as follows.*
(*i*) *If f is ϕ_t then rank$(f) = 0$.*
(*ii*) *If f is $\psi(f_1, \ldots, f_n)$ then*

$$rank(f) = \max\{rank(f_i) : i = 1, \ldots, n\}.$$

(*iii*) *If f is $E[g|t]$ then rank$(f) = rank(g) + 1$.*
(*b*) *We say that two processes x and y have the same adapted distribution up to rank n if for every f with rank$(f) \le n$, $E[f(x)] = E[f(y)]$. This new relation is denoted by $x \equiv_n y$.*

Observe that this definition remains unchanged if we deal with *CP*'s instead of *AF*'s. We can now answer the question posed before the previous definition.

$x \equiv_0 y$ if and only if x and y have the same finite dimensional distribution. Thus the relation $x \equiv_0 y$ is nothing new. (Check it!).

The next proposition collects some technical facts that we will need later. Its proof is left as an exercise.

PROPOSITION 1D.2. (*a*) *If x is a **version** of y (that is, for all t, $x(\omega, t) = y(\omega, t)$ a.s.) then $x \equiv y$.*
(*b*) *If $x \equiv_0 y$ and x is uniformly integrable then so is y.*
(*c*) *If M, N are Polish spaces, x and y have values in M, $x \equiv y$ and $h : M \to N$ is a Borel function, then $h(x) \equiv h(y)$, where the process $h(x)$ is defined in the natural way by $(h(x))_t = h(x_t)$.* ⊣

The following proposition relates adapted equivalence to Markov processes.

PROPOSITION 1D.3. (*a*) *If $x \equiv_1 y$ and x is Markov then y is also Markov.*
(*b*) *If x and y are Markov then the relations $x \equiv_0 y$, $x \equiv_1 y$, and $x \equiv y$ are equivalent.*

PROOF. (a) This result follows from Proposition 1D.2 (c) and the fact that for any random variable z which is measurable with respect to $\sigma(x_s)$, there is a Borel function $h : \mathbb{R} \to \mathbb{R}$ such that $z = h(x_s)$.

(b) The proof can be found in Hoover and Keisler [1984], Theorem 2.8.   ⊣

Part (b) of this proposition explains a key feature of Markov processes. All the information that is needed is already contained in the process. This observation allows us to give an exercise at this point that deals with the information contained in the process itself. We will study this sort of situation in full generality later on in Chapter 6.

EXERCISE 1D.4. *Let x be an adapted process defined on Ω and let $\mathcal{H}_t = \sigma(x_s : s \leq t)$ be its natural filtration. Clearly. for every t we have $\mathcal{H}_t \subseteq \mathcal{F}_t$. Let Ω^x be the adapted space obtained by replacing the filtration $(\mathcal{F}_t)$ by $\mathcal{H}_t^+$ (i.e., the smallest right continuous filtration that satisfies the usual conditions and contains $\mathcal{H}_t$). Find conditions that guarantee that*

$$(\Omega, x) \equiv_1 (\Omega^x, x).$$

The relation $\equiv_1$ was first defined in the unpublished paper Aldous [1981]. Aldous called the relation **synonymity** and suggested that it was good enough to capture the idea of two processes having the same probabilistic behavior in the context of the general theory of processes. Here are his words (see Aldous [1981], page 14):

"We generally the attitude that synonymous processes are really the same. To justify this, we need to be sure that any property of processes which may be of interest is invariant under synonymity."

Is he right? Observe that all the results where we have used $\equiv$ remain true when $\equiv$ is replaced by $\equiv_1$. Here is a proposition due to Hoover and Keisler [1984] that tells us that $\equiv$ and $\equiv_1$ are not the same. Later we will show that, in a certain sense, the relation $\equiv$ fulfills the requirements that Aldous expected from $\equiv_1$.

PROPOSITION 1D.5. *For each n, there are processes x and y such that $x \equiv_n y$ but not $x \equiv_{n+1} y$.*

PROOF. Here we just give the proof in the case $n = 0$. The proof is by induction, building on this case. Consider the space $\Omega = \{-1, 1\}^{\{1/4.1/2\}}$ with the counting measure on its power set, identifying an ω with the pair $(\omega(1/4), \omega(1/2))$. Consider the following filtration on Ω.

$$\mathcal{F}_t = \{\emptyset, \Omega\} \qquad\qquad \text{for } 0 \leq t < \tfrac{1}{4},$$

$$\mathcal{F}_t = \{\{(-1, 1), (-1, -1)\}, \{(1, 1), (1, -1)\}, \emptyset, \Omega\} \text{ for } \tfrac{1}{4} \leq t < \tfrac{1}{2},$$

$$\mathcal{F}_t = \wp(\Omega) \qquad\qquad \text{for } \tfrac{1}{2} \leq t.$$

Now consider the processes x and y defined by:

$$x(\omega, t) = y(\omega, t) = 0 \qquad \text{for } 0 \leq t < \tfrac{1}{4},$$

$$x(\omega, t) = \omega(\tfrac{1}{4}) \qquad\qquad \text{for } \tfrac{1}{4} \leq t,$$

$$\text{and } y(\omega, t) = \omega(\tfrac{1}{2}) \qquad \text{for } \tfrac{1}{4} \leq t.$$

Observe that $y(\omega, t) = x(h(\omega), t)$ for each t, where the function $h : \Omega \to \Omega$ is defined by

$$h((\omega(1/4), (\omega(1/2)))) = (\omega(1/2), \omega(1/4))).$$

This h is obviously a bijection of Ω, and since P is the counting measure it is measure preserving. Consequently, x and y have the same finite dimensional distribution, that is, $x \equiv_0 y$.

In order to see that not $x \equiv_1 y$, it is enough to find a property expressible by means of adapted functions which holds for one process and not for the other. In this case it is easy. $x_{1/4}$ is $\mathcal{F}_{1/4}$-measurable but $y_{1/4}$ is not. We leave it to the reader to verify this fact. A description of this example by means of a betting scheme was given by the second author (see López [1989]).   ⊣

This proposition tells us that $\equiv$ is strictly stronger than $\equiv_1$ and all the other $\equiv_n$'s. However, one could object to this answer because it uses spaces and processes which were built in an ad-hoc manner for the problem, and do not seem to show up naturally in any probability textbook. It would be more interesting if we could come up with a known property of stochastic processes which is preserved by $\equiv$ but not by $\equiv_1$.

In general, the *preservation problem* asks whether a given property is preserved under $\equiv$ or $\equiv_n$. We have a good deal of work ahead of us just to look at some of the most common properties of stochastic processes. Here are some sample results along this line from the literature. (For definitions see Dellacherie and Meyer [1978] or Ethier and Kurtz [1986]).

PROPOSITION 1D.6. (a) *If* $\gamma \equiv_1 \sigma$ *and* γ *is a stopping time then* σ *is a stopping time.*

(b) *If* $x \equiv_1 y$ *and* x *is a local martingale then* y *is a local martingale.*

(c) *If* $x \equiv_1 y$ *and* x *is a semimartingale then* y *is a semimartingale.*

PROOF. (a) Observe that γ is a stopping time if and only if for every t, $\min(\gamma, t)$ is $\mathcal{F}_t$-measurable. In this form the property can be expressed using adapted functions.

(b) and (c) were obtained in the paper Hoover [1984]. These results were consequences of theorems that were originally proved for other purposes. The arguments presented there depend on an advanced knowledge of the general theory of processes and will not be studied here.   ⊣

Part (b) of this proposition lets us point out an interesting fact. The definition of a local martingale (see Dellacherie and Meyer [1988] or Ethier and Kurtz [1986]) involves the existence of a localizing sequence of stopping times. Let us define this interesting concept. In what follows we use the expression "probabilistic property" in a lax, intuitive sense. For example, being a martingale is a probabilistic property.

DEFINITION 1D.7. *Let* $\mathbb{P}$ *be a property of stochastic processes. A stochastic process* x *is a* **local** $\mathbb{P}$ *if there is an increasing sequence of stopping times* $(\sigma_n)_{n \in \mathbb{N}}$, *such that:*

(i) $\lim \sigma_n = 1$ *a.s.*

(ii) *For each n, the stopped process x^{σ_n}, defined by*

$$x^{\sigma_n}(\omega, t) = x(\min(\sigma_n(\omega), t), t),$$

has property $\mathbb{P}$. *We call the sequence* (σ_n) *a* $\mathbb{P}$-**localizing sequence** *for x.*

Observe that the existence of a $\mathbb{P}$-localizing sequence is a type of statement that is not expressible by means of the language of adapted functions. But nevertheless a notion such as a local martingale is preserved under $\equiv_1$. This observation is a good reason for posing the following exercises.

EXERCISES 1D.8. (a) *Suppose a property* $\mathbb{P}$ *of stochastic processes is preserved under* $\equiv$ *or* $\equiv_1$. *Is the property* local($\mathbb{P}$) *preserved under* $\equiv$ *or* $\equiv_1$? *Can we identify properties* $\mathbb{P}$ *with a particular form for which* local($\mathbb{P}$) *is preserved under* $\equiv$ *or* $\equiv_1$? *Is there a general result, something like a quantifier elimination theorem for stopping times, that answers these questions?*

(b) *Let γ and σ be stopping times. Does* $(x, \gamma) \equiv_1 (y, \sigma)$ *imply* $x^\gamma \equiv_1 y^\sigma$? *Does* $(x^\gamma, \gamma) \equiv_1 (y^\sigma, \sigma)$ *imply* $x \equiv_1 y$? *Feel free to impose regularity conditions on the paths of the processes. Is there any difference if instead of* $\equiv_1$ *we use* $\equiv$?

(c) *Suppose $x \equiv y$ (or $x \equiv_1 y$) and there exists a stopping time θ such that x^θ is a martingale. Does there exist a stopping time σ such that y^σ is also a martingale?*

(d) *Which of the following properties are preserved under* $\equiv_1$ *or* $\equiv$? *See Dellacherie and Meyer* [1978] *and Dellacherie and Meyer* [1988] *for definitions of unknown concepts.*

Quasimartingale
Class D process
Quasi-left continuity
Strong Markov
Strong martingale
Amart

Observe that all the notions in (d) involve stopping times in their definitions. What is interesting is that in their formulations the statements are "universal" in the sense that they involve all the stopping times defined on the adapted space. This feature is different from existential notions such as local martingale. It is likely that the notions in (d) are not even preserved by the adapted equivalence relation $\equiv$. At any rate we should get an interesting lesson from whatever answer we get.

1E. Universal spaces

We will first introduce the notion of a universal probability space, and then extend the notion to adapted spaces by replacing the distribution by the adapted distribution. Using familiar results from basic measure theory, one can easily construct probability spaces which are universal in the sense that every probability distribution is the law of some random variable on the space. In the literature the name universal probability space is not commonly used. Instead one uses the notion of an atomless probability space, which turns out to be equivalent to universality.

However, the situation is completely different for adapted spaces. We will show that none of the ordinary adapted spaces from the classical literature are universal. This will set the stage for the next chapter, where universal adapted spaces will be constructed by nonstandard methods.

DEFINITION 1E.1. *A probability space Ω is **universal** if given any Polish space M and M-valued random variable y on an arbitrary probability space Γ, there is an M-valued random variable x on Ω such that $x \equiv_0 y$.*

*A probability space $(\Omega, \mathcal{F}, P)$ is **atomless** if for every measurable set $A \in \mathcal{F}$ with $P[A] > 0$ there exists $B \in \mathcal{F}$ with $B \subset A$ such that $0 < P[B] < P[A]$.*

The following theorem explains why we want to have atomless spaces.

THEOREM 1E.2. *A probability space is universal if and only if it is atomless.*

PROOF. We give a hint and leave the details as an exercise. (For a detailed proof see Billingsley [1979]). Assume Ω is atomless. Given y on Γ, choose a sequence y_n of simple random variables on Γ converging to y, construct a sequence x_n on Ω with the same joint distribution, and take the limit in probability. ⊣

One of the simplest examples of an atomless probability space is the real unit interval $[0, 1]$ with the Borel σ-algebra and Lebesgue measure, which we denote by $([0, 1], \mathcal{B}, \lambda)$. Thus it could be said that with respect to random variables and their laws we can do everything inside the unit interval.

Our next theorem will show that atomless probability spaces are universal for stochastic processes as well as for random variables.

We first prove a lemma which holds for adapted spaces as well as probability spaces. It will give us a way of showing that two stochastic processes are (adapted) equivalent by showing that certain random variables or processes with nice paths are (adapted) equivalent. Throughout this book we will use the notation $\mathbb{I}_A$ for the characteristic function (or indicator function) of a set A. A process is said to be **continuous** if at every ω its path is continuous as a function defined on $[0, 1]$. Similarly, a process is said to be right continuous with left limits, or **cadlag**, if at every ω its path is right continuous with left limits as a function defined on $[0, 1]$.

LEMMA 1E.3. (*a*) *Let x be a stochastic process on Ω with values in M. There exists a sequence (t_n) from $[0, 1]$, and Borel functions $\psi : [0, 1] \times M^{\mathbb{N}} \to 2^{\mathbb{N}}$ and $\phi : 2^{\mathbb{N}} \to M$, such that for all $t \in [0, 1]$, $x(\omega, t) = \phi(\psi(t, \langle x(\omega, t_n) \rangle))$ a.s.*

(*b*) *If x is a stochastic process in an adapted space Ω, there exists a cadlag process x' and a Borel function ϕ such that $\phi(x'(\omega, t)) = x(\omega, t)$. Moreover, if $x \equiv y$ we can choose ϕ and cadlag processes z and w such that $z \equiv w$, $x = \phi(z)$ a.s., and $y = \phi(w)$ a.s.*

PROOF. We first need to prove the following fact:

x has a $\sigma(x_t : t \leq 1) \times \mathcal{B}$-measurable version $\hat{x}$, where $\mathcal{B}$ is the Borel σ-algebra on $[0, 1]$.

In order to get this result, we show that for every $\mathcal{F}_1 \times \mathcal{B}$-measurable set S, there exists a $\sigma(x_t : t \leq 1) \times \mathcal{B}$-measurable process e such that for each t.

$$e_t = E[S(t)|\sigma(x_t : t \leq 1)] \, a.s.$$

As usual, this is done with a monotone class argument. Here is the first step. If S is a measurable rectangle $A \times B$ then the function

$$e(t, \omega) = \mathbb{I}_B(t) E[\mathbb{I}_A|\sigma(x_t : t \leq 1](\omega)$$

is $\sigma(x_t : t \leq 1) \times \mathcal{B}$-measurable, since it is the product of two variables which are $\sigma(x_t : t \leq 1) \times \mathcal{B}$-measurable. Then for each t.

$$e_t = E[S(t)|\sigma(x_t : t \leq 1)] \, a.s.$$

as required.

Now consider a set S of the form $\{x \in U\}$ where U is an open set in M. Then for each t. $S(t)$ is $\sigma(x_t : t \leq 1)$-measurable, and so for the process e corresponding to S from the previous paragraph. $e_t = S(t) \, a.s.$ This fact, together with the separability of M, lets us conclude that x has a version $\hat{x}$ which is $\sigma(x_t : t \leq 1) \times \mathcal{B}$-measurable. Now let us work with $\hat{x}$.

We can find a countable sequence (t_n) in $[0, 1]$ such that $\hat{x}$ is $\sigma(x_{t_n} : n \in \mathbb{N}) \times \mathcal{B}$-measurable. If $\{U_k\}$ is a countable basis for M, $E_{nk} = \{x_{t_n} \in U_k\}$, and $\{[a_n, b_n)\}$ is an enumeration of the right open intervals with rational endpoints, let $\{[a_m, b_m) \times F_m : m \in \mathbb{N}\}$ be an enumeration of the family of products between the $[a_n, b_n)$'s and the E_{nk}'s. Now define $z : \Omega \times [0, 1] \to \{0, 1\}^{\mathbb{N}}$ by

$$z(\omega, t) = (\mathbb{I}_{[a_m, b_m) \times F_m}(\omega, t))_{m \in \mathbb{N}}.$$

Clearly z is cadlag. $\hat{x}$ is $\sigma(z)$-measurable and so there exists a Borel function $\phi : \{0, 1\}^{\mathbb{N}} \to M$ such that $\hat{x} = \phi(z)$.

If we write z carefully we can see that it has the form required in (a):

$$z(\omega, t) = (\mathbb{I}_{[a_m, b_m)}(t) \cdot \mathbb{I}_{F_m}(\omega))_{m \in \mathbb{N}} = (\mathbb{I}_{[a_m, b_m)}(t) \cdot \mathbb{I}_{\{x_{t_m} \in U_m\}}(\omega))_{m \in \mathbb{N}} =$$

$$(\mathbb{I}_{[a_m, b_m)}(t) \cdot \mathbb{I}_{U_m}(x_{t_m}))_{m \in \mathbb{N}} = \psi(t, \langle (x_{t_m}) : m \in \mathbb{N} \rangle).$$

We leave it to the reader to verify that defining w by $\psi(t, \langle (y_{t_m}) : m \in \mathbb{N} \rangle)$ we obtain (b). $\dashv$

As an application, we show that atomless probability spaces are universal for stochastic processes with respect to finite dimensional distribution (see Keisler [1988] Proposition 3.2).

COROLLARY 1E.4. *Let Ω be an atomless probability space. Then for every stochastic process x on a probability space Γ there is a stochastic process y on Ω such that $x \equiv_0 y$.*

PROOF. Let ϕ, ψ, and (t_n) be as in part (a) of the preceding proposition. Since Ω is a universal probability space, there is a random variable z on Ω with values in $M^{\mathbb{N}}$ which has the same law as $(x(t_n))$. Since $x(\omega, t) = \phi(\psi(t, \langle x(\omega, t_n) \rangle))$, the stochastic process $y(\omega, t) = \phi(\psi(t, z))$ on Ω has the same finite dimensional distribution as x. $\dashv$

Let us now introduce the analogous notion of universality for adapted spaces and adapted distribution.

DEFINITION 1E.5. *An adapted space Ω is* **universal** *(or $\equiv_n$-universal) if given any Polish space M and M-valued stochastic process y on an arbitrary adapted space Γ, there is an M-valued stochastic process x on Ω such that $x \equiv y$ (or $x \equiv_n y$).*

An adapted space is **universal for random variables** *if the condition in the preceding paragraph holds when x, y are taken to be random variables with values in any Polish space M.*

It is clear that every universal adapted space is $\equiv_n$-universal for each n, and every $\equiv_{n+1}$-universal adapted space is $\equiv_n$-universal.

In an adapted space, an argument can sometimes be simplified by restricting our attention to random variables rather than considering arbitrary stochastic processes. As another application of Lemma 1E.3, we show that the notion of a universal adapted space permits this kind of simplification.

LEMMA 1E.6. *An adapted space is universal if and only if it is universal for random variables.*

PROOF. We first prove the easy direction. Suppose Ω is universal, and let x be a random variable on some adapted space. Let $\bar{x}$ be any stochastic process such that $\bar{x}_0 = x$. Then there is a stochastic process $\bar{y}$ on Ω with $\bar{y} \equiv \bar{x}$. Let y be the random variable $\bar{y}_0$ on Ω. Then $y \equiv x$, and hence Ω is universal for random variables.

Now suppose Ω is universal for random variables, and let x be a stochastic process on some adapted space. We use Lemma 1E.3 to obtain a sequence $(t_n)_{n \in \mathbb{N}}$ of elements in $[0, 1]$ and two Borel functions $\psi : [0, 1] \to 2^{\mathbb{N}}, \phi : 2^{\mathbb{N}} \to M$ such that for all $t \in [0, 1]$,

$$x(\cdot, t) = \phi(\psi(t, \langle x(\cdot, t_n) : n \in \mathbb{N}\rangle)) \ a.s.$$

Since Ω is universal for random variables, there is a random variable $z : \Omega \to M^{\mathbb{N}}$ such that

$$z(\cdot) \equiv \langle x(\cdot, t_n) : n \in \mathbb{N}\rangle.$$

Then

$$(t, z(\cdot)) \equiv (t, \langle x(\cdot, t_n) : n \in \mathbb{N}\rangle).$$

Thus

$$\phi(\psi(t, z(\cdot))) \equiv \phi(\psi(t, \langle x(\cdot, t_n) : n \in \mathbb{N}\rangle)),$$

by Proposition 1D.2 (c), because ϕ and ψ are Borel functions. But the last function is a version of x, so if we define

$$y(\cdot, t) = \phi(\psi(t, z(\cdot))),$$

we have $y \equiv x$.   ⊣

The next result is just a restatement of Corollary 1E.4.

COROLLARY 1E.7. *Every adapted space which is atomless as a probability space is $\equiv_0$-universal.*   ⊣

Our next result is negative—it shows that no adapted space of the kind that is commonly found in the literature is universal. We first make the notion of an "ordinary" space precise.

DEFINITION 1E.8.   *By an* **ordinary probability space** *we will mean a probability space of the form* $(\Omega, \mathcal{F}, \mu)$ *where* Ω *is a Polish space and* μ *is the completion of a probability measure on the family of Borel sets in* Ω.

By an **ordinary adapted space** *we simply mean an adapted space which is ordinary as a probability space.*

In the literature, one usually works with adapted spaces which are ordinary in the above sense.

EXAMPLE 1E.9.   *Each of the following commonly used adapted spaces is ordinary. In each case, M is a Polish space, the probability measure is Borel, and the filtration is the natural one.*

(1) *The space $C([0, 1], M)$ of continuous paths with the metric of uniform convergence.*

(2) *The space $D([0, 1], M)$ of cadlag paths with the Skorokhod topology.*

(3) *The space $L^0([0, 1], M)$ of Lebesgue measurable paths with the metric of convergence in measure.*

(4) *The space $L^p([0, 1], M)$ of p-integrable paths with the metric of convergence in p-th mean.*

By analogy with the situation for probability spaces, one might expect that there are ordinary adapted spaces which are universal. However, this is not the case. We conclude this chapter by showing that ordinary adapted spaces are *never* universal.

A set A is said to be **independent** of a σ-algebra $\mathcal{H}$ if A is independent of each set $B \in \mathcal{H}$. Equivalently, the conditional expectation $E[A|\mathcal{H}]$ is constant almost surely. An **independent family** is a family of sets such that each set in the family has measure in $(0, 1)$ and is independent of every other set in the family.

LEMMA 1E.10.   *In an ordinary probability space, every independent family of sets is at most countable.*

PROOF. Suppose that there is an uncountable independent family $\mathcal{C}$. We may assume without loss of generality that for some $r > 0$, each set in $\mathcal{C}$ has measure in $(r, 1-r)$. Then $P[A \Delta B] \geq r/2$ for each distinct $A, B \in \mathcal{C}$. M has a countable open basis, which generates a countable Boolean algebra of sets $\mathcal{X}$. For each $A \in \mathcal{C}$ there exists $X_A \in \mathcal{X}$ such that $P[A \Delta X_A] < r/4$. Since $\mathcal{C}$ is uncountable, there are distinct $A, B \in \mathcal{C}$ with $X = X_A = X_B$. But then $P[A \Delta B] \leq P[A \Delta X] + P[B \Delta X] < r/2$, contradiction.   ⊣

THEOREM 1E.11.   *No ordinary adapted space is universal. In fact, no ordinary adapted space is even $\equiv_1$-universal.*

PROOF. Suppose Ω is an $\equiv_1$-universal adapted space. The idea is to get an independent family of sets in Ω by approaching each $\mathcal{G}_u$ from the left. For each time $u \in (0, 1]$, there is an adapted space $(\Gamma^u, \mathcal{G}_t^u, Q^u)$ with a set $A^u \in \mathcal{G}_u^u$ such that $Q^u[A^u] = 1/2$ and A^u is independent of $\mathcal{G}^u_t$ for all $t < u$. Let x^u be the characteristic function of A^u. Then for each u there must be a random variable y^u on Ω such that $y^u \equiv_1 x^u$. We have

$$E[x^u|\mathcal{G}^u_u] = x^u \in \{0, 1\} \, a.s.,$$

and

$$E[x^u|\mathcal{G}^u_t] = 1/2 \, a.s. \text{ for each } t < u.$$

Then

$$E[y^u|\mathcal{F}_u] = y^u \in \{0, 1\} \, a.s.,$$

and

$$E[y^u|\mathcal{F}_t] = 1/2 \, a.s. \text{ for each } t < u.$$

It follows that y^u is the indicator function of a set B^u for each u, and $\{B^u : u \in (0, 1]\}$ is an uncountable independent family of sets in Ω. Thus by Lemma 1E.10, Ω is not an ordinary adapted space. ⊣

We now have all the basic tools needed for the model theory of stochastic processes, and move on to the next chapter.

CHAPTER 2

HYPERFINITE ADAPTED SPACES

In this chapter we study a family of adapted spaces that has been widely and successfully used in the nonstandard approach to stochastic analysis, the hyperfinite adapted spaces. The results in the monograph "An Infinitesimal Approach to Stochastic Analysis", Keisler [1984], prompted a natural question: Why are these spaces so "rich" or "well behaved"?

In order to answer this question, we built a probability logic adequate for the study stochastic processes: adapted probability logic (see Keisler [1979], Keisler [1985], Keisler [1986a], Keisler [1986b], Hoover and Keisler [1984], Fajardo [1985a]). This is the origin of the theory we are describing in this book. We chose a somewhat different approach in Chapter 1 in order to introduce the theory in a smooth way without any need for a background in logic.

Basic nonstandard probability theory is a necessary prerequisite for most of this chapter. This theory is readily available to the interested mathematician without going through the technical literature on nonstandard analysis (see, among others, Albeverio, Fenstad, Hoegh-Krohn, and Lindstrøm [1986], Cutland [1983], Fajardo [1990b], Lindstrøm [1988], Stroyan and Bayod [1986] and Keisler [1988]). Nonetheless, in the following section we collect the main definitions and results needed in this book.

This seems to be an appropriate place to add a remark about the use of non-standard analysis. It has been very hard to convince the mathematical community of the attractive characteristics of nonstandard analysis and its possible uses as a mathematical tool. This chapter, among other things, continues the task of showing with direct evidence the enormous potential that we believe nonstandard analysis has to offer to mathematics. The paper Keisler [1994] examines some of the reasons why nonstandard analysis has developed in the way we know it today and discusses the perspectives and possibilities in the years to come.

2A. Basic results

In this section we will browse through the definitions and results most needed from the nonstandard probability literature. It is by no means self-contained, and the reader will already have to be acquainted with the basic ideas and methods of the nonstandard theory. We follow the notation from Keisler [1988].

CONVENTION 2A.1. *Throughout this book, whenever we use notions from nonstandard probability theory it will be understood that we are working in nonstandard universe which satisfies the* **Saturation Principle,** *that any decreasing chain of nonempty internal sets has a nonempty intersection.*

This convention is required for the Loeb measure construction. A nonstandard universe which satisfies the Saturation Principle is said to be ω_1-saturated. On a few rare occasions, we will need to assume that the nonstandard universe is κ-saturated for some larger cardinal κ.

DEFINITION 2A.2. *A* **Loeb probability space** *is a probability space* $L(\Omega) = (\Omega, L(\mathcal{A}), L(\mu))$ *where* $(\Omega, \mathcal{A}, \mu)$ *is an internal *finitely additive probability space, $L(\mathcal{A})$ is the Loeb σ-algebra over $\mathcal{A}$, and $L(\mu)$ is the Loeb measure associated with μ.*

A **hyperfinite probability space** *is a Loeb probability space* $L(\Omega) = (\Omega, L(\mathcal{A}), L(\mu))$ *where* Ω *is a hyperfinite set,* $\mathcal{A}$ *is the algebra of internal subsets of* Ω*, and* μ *is the internal counting measure over* Ω*. The hyperfinite size (i.e., the internal cardinality) of a hyperfinite set A is denoted by $|A|$.*

Using the Saturation Principle. it is easy to show that every atomless Loeb probability space has an uncountable family $\mathcal{C}$ of measurable sets such that $L(\mu)[A\Delta B] = 1/4$ for all distinct $A, B \in \mathcal{C}$. Thus by Lemma 1E.10, no atomless Loeb space is ordinary in the sense of Definition 1E.8, and its measure algebra cannot even be imbedded in that of an ordinary probability space. The papers Jin and Keisler [2000] and Keisler and Sun [2001] give further results which compare Loeb spaces to classical probability spaces.

We are now ready to define the hyperfinite adapted spaces.

DEFINITION 2A.3. *(a) Let H be a positive infinite hyperinteger which will remain fixed throughout this chapter. The* **hyperfinite time line** $\mathbb{T}$ *of mesh $H!$ is the set of points in *$[0, 1]$ of the form $n/H!$ with $n < H!$. Note that in $\mathbb{T}$ we can find all the rational numbers in $[0, 1]$.*

(b) Let Ω be a hyperfinite set of the form $\Omega_0^{\mathbb{T}}$ (the internal set of all internal functions from $\mathbb{T}$ into Ω_0) where Ω_0 is a finite or hyperfinite set with at least two elements. The internal hyperfinite adapted space over Ω is the internal structure $(\Omega, (\mathcal{A}_t)_{t \in \mathbb{T}}, \mu)$ where μ is the uniform counting measure and for each $t \in \mathbb{T}$, $\mathcal{A}_t$ is the algebra of all internal sets closed under the internal equivalence relation $\approx_t$ defined as follows: $\omega \approx_t \omega'$ if and only if for each $s \leq t, \omega(s) = \omega'(s)$. The equivalence class of an $\omega \in \Omega$ under this relation is denoted by $(\omega|t)$. Observe that $\mathcal{A}_1$ is the algebra of internal subsets of Ω.

Sometimes it is convenient to extend the internal filtration $\mathcal{A}_t$ to all hyperreal times $t \in *[0, 1]$ by defining

$$\mathcal{A}_t = \bigcap \{\mathcal{A}_s : t \leq s \in \mathbb{T}\}.$$

(c) *The* **hyperfinite adapted space** *over* Ω *is the adapted space*

$$\Omega = (\Omega, \mathcal{F}_t, L(\mu))$$

where for each $t \in [0, 1]$, $\mathcal{F}_t$ *is the completion of the σ-algebra of all Loeb measurable sets closed under the external equivalence relation* $\sim_t$ *defined as follows:* $\omega \sim_t \omega'$ *if and only if for all* $s \in \mathbb{T}$ *with* $^o s = t, \omega \approx_s \omega'$.

The hyperfinite adapted spaces introduced in part (c) are just particular cases of a more general construction that we will occasionally need. Given an arbitrary internal adapted space $(\Omega, (\mathcal{B}_t)_{t \in {}^*[0,1]}, \delta)$, the corresponding **Loeb adapted space** is the adapted space $L(\Omega) = (\Omega, (\mathcal{G}_t)_{t \in [0,1]}, L(\delta))$ with the filtration

$$\mathcal{G}_t = \sigma((\bigcup_{s \approx t} \mathcal{B}_s) \vee \mathcal{N}),$$

where $\mathcal{N}$ is the set of null sets of $L(\delta)$. It can be shown that the filtration $(\mathcal{G}_t)$ is always right continuous, that is, $\mathcal{G}_t = \bigcap_{s > t} \mathcal{G}_s$ for all $t \in [0, 1)$. In particular, given a hyperfinite set $\Omega = \Omega_0^{\mathbb{T}}$, the Loeb adapted space corresponding to the internal hyperfinite adapted space over Ω is the hyperfinite adapted space over Ω as defined in (c).

In the books Albeverio, Fenstad, Hoegh-Krohn, and Lindstrøm [1986], Stroyan and Bayod [1986] and the papers Cutland [1983], Keisler [1984] and Peña [1993] one can find a very detailed study of the main properties of the spaces introduced in this definition.

CONVENTION 2A.4. . *For the rest of this chapter, it will always be understood that Ω is the hyperfinite adapted space constructed from Ω_0 and the hyperfinite time line $\mathbb{T}$ as in Definition 2A.3.*

When working with hyperfinite adapted spaces, Nonstandard practitioners continually make use of a technique, called the "lifting procedure", which at the heart of nonstandard analysis' success. It can be intuitively explained as follows.

The general idea of the lifting procedure is to translate problems about continuous mathematical objects into problems about discrete objects in the nonstandard universe. The first step is to introduce hyperfinite counterparts to the standard probability notions under consideration. Next, solve a hyperfinite counterpart of the given problem using the tools of nonstandard analysis. Finally, going back to the standard world, take a "standard part" of the hyperfinite solution to get a solution of the original problem. For a detailed account of this process see the references given above. The paper Keisler [1991] presents another view of the transition from discrete to continuous time. These ideas and others will be fully discussed in Chapters 7 and 8.

Here we are going to follow the lifting procedure. The next definition is the starting point.

DEFINITION 2A.5. (a) *Let x be a function from Ω into $\mathbb{R}$. A* **lifting** *of x is an internal function $X : \Omega \to {}^*\mathbb{R}$ such that*

$$^o X(\omega) = x(\omega) \ a.s. \ in \ L(\Omega).$$

(b) Let x be a function from $\Omega \times [0, 1]$ into $\mathbb{R}$. A lifting of x is an internal function $X : \Omega \times \mathbb{T} \to {}^*\mathbb{R}$ such that

$$^{o}X(\omega, t) = x(\omega. {}^{o}t) \ a.s. \ in \ L(\Omega \times \mathbb{T}).$$

(c) Given an internal $X : \Omega \to {}^*\mathbb{R}$ and one of our internal algebras $\mathcal{A}_t$, the internal conditional expectation of X with respect to $\mathcal{A}_t$ is the defined by the formula

$$E[X|\mathcal{A}_t](\omega) = \sum_{\omega' \in (\omega|t)} X(\omega')/|(\omega|t)|.$$

The space $L(\Omega \times \mathbb{T})$ in part (b) is the hyperfinite probability space over the hyperfinite set $\Omega \times \mathbb{T}$ (with the usual hyperfinite counting measure). For more about products of Loeb spaces see Keisler [1988]. Notice how the internal hyperfinite notion in part (c) captures the usual concept of conditional expectation. This definition can also be extended in a natural way to arbitrary internal algebras over Ω.

DEFINITION 2A.6. *A stochastic process x is said to be* **right continuous in probability** *if for every $t \in [0, 1)$ and every real $\varepsilon > 0$, the limit as $u \downarrow t$ of the probability that $x(\cdot, u)$ is within ε of $x(\cdot, t)$ is one.*

This class of processes plays an important role in our theory. Here is the hyperfinite counterpart of this notion.

DEFINITION 2A.7. *An internal stochastic process X is said to be a* **right lifting** *of a stochastic process x if for each $t \in [0. 1)$ there is an element $s \in \mathbb{T}$, with $s \approx t$, such that $X(\cdot, u)$ is a lifting of $x(\cdot, t)$ whenever $u \in \mathbb{T}$ with $u \approx t$ and $u \geq s$.*

The following theorem collects all relevant results involving the notions just defined. We omit the proof and refer the reader to Keisler [1988].

THEOREM 2A.8. *Let $(\Omega. \mathcal{F}_t, L(\mu))$ be a hyperfinite adapted space.*
(a) The filtration $(\mathcal{F}_t)$ is right continuous, that is, for all $t \in [0, 1)$,

$$\mathcal{F}_t = \bigcap_{s > t} \mathcal{F}_s.$$

(b) A function $x : \Omega \to \mathbb{R}$ is $L(\Omega)$-measurable if and only if it has a lifting.
(c) A function $x : \Omega \times [0, 1] \to \mathbb{R}$ has a lifting if and only if there is a stochastic process $\hat{x} : \Omega \times [0, 1] \to \mathbb{R}$ such that $x = \hat{x}$ a.s with respect to $L(\Omega) \times \beta$, where β is the Borel measure on $[0, 1]$.
(d) Let x be a bounded random variable on Ω. If X is a bounded lifting of x, then for any $t \in \mathbb{T}$. the internal function $E[X|\mathcal{A}_t]$ is a lifting of $E[x|L(\mathcal{A}_t)]$.
(e) If $x \equiv_0 y$ and x is right continuous in probability then so is y.
(f) A stochastic process is right continuous in probability if and only if it has a right lifting.
(g) Given a bounded random variable x on Ω. with a bounded lifting X, the internal stochastic process $Y(\omega, t) = E[X|\mathcal{A}_t](\omega)$ is a right lifting of the stochastic process $x(\omega, s) = E[x|\mathcal{F}_s](\omega)$. ⊣

2B. Adapted lifting theorem

We now return to adapted distributions. Following the ideas of the previous section, the first thing to do is to introduce a hyperfinite counterpart to the notion of the value of a conditional process (or adapted function). We will not bother to write down the formal definitions, but will simply explain the changes that are needed.

Given an n-fold conditional process f and an internal stochastic process X, the value $f(X)$ will be an n-fold internal stochastic process. It is defined by making the following changes to Definitions 1C.1 and 1C.3. In the Basis and Composition Steps, use $^*\phi$ and $^*\psi$ in place of ϕ and ψ and use times from $\mathbb{T}$ instead of $[0, 1]$. In the Conditional Expectation Step, instead of conditioning with respect to $\mathcal{F}_t$, take the internal conditional expectation with respect to $\mathcal{A}_t$. The rest remains unchanged.

The following theorem from Keisler [1988] is crucial for our theory, so much so that it deserves a special section for itself. It extends the results about liftings of stochastic processes to liftings of conditional processes.

THEOREM 2B.1. (*Adapted Lifting Theorem*) *Let x be a stochastic process on Ω and X a lifting of x. Then for each conditional process f, its value $f(X)$ on X is a lifting of its value $f(x)$ on x. Moreover, if X is a right lifting of x then each $f(X)$ is a right lifting of $f(x)$.*

PROOF. We proceed by induction. The Basis and Composition Steps are immediate, using the Transfer Principle. The Conditional Expectation Step requires a little more attention. Let $f \in CP$ and for simplicity assume that f has only one time variable t. Now let $g(s, t)$ be $E[f|s]$. When interpreting we have:

$$f(x) : \Omega \times [0, 1] \to \mathbb{R},$$

$$g(x) : \Omega \times [0, 1]^2 \to \mathbb{R},$$

$$f(X) : \Omega \times \mathbb{T} \to {}^*\mathbb{R},$$

and

$$g(X) : \Omega \times \mathbb{T}^2 \to {}^*\mathbb{R}.$$

By the induction hypothesis, $f(X)$ is a lifting of $f(x)$, which means that the set

$$\{(\omega, t) \in \Omega \times \mathbb{T} : {}^o f(X)(\omega, t) = f(x)(\omega, {}^o t)\}$$

has measure one on $L(\Omega \times \mathbb{T})$. We want to show that the set

$$B = \{(\omega, t, s) \in \Omega \times \mathbb{T}^2 : {}^o g(X)(\omega, t, s) = g(X)(\omega, {}^o t, s)\}$$

has measure one on $L(\Omega \times \mathbb{T}^2)$. From the induction hypothesis and the Keisler-Fubini theorem, the set

$$A = \{t \in \mathbb{T} : f(X)(t) \text{ lifts } f(x)({}^o t)\}$$

has measure one on $L(\mathbb{T})$.

Now for each $t \in A$, form the martingale

$$g(x)(^{o}t, r) = E[f(x)(^{o}t)|\mathcal{F}_r],$$

which has the right lifting

$$g(X)(t, s) = E[f(X)(t)|\mathcal{A}_s]$$

(see Keisler [1988], Theorem 6.2.). Since every right lifting is a lifting, $g(X)(t, s)$ is a lifting of $g(x)(^{o}t, r)$. This means that the set

$$A_t = \{(\omega, s) \in \Omega \times \mathbb{T} : {}^{o}g(X)(\omega, t, s) = g(X)(\omega, {}^{o}t, s)\}$$

has measure one on $L(\Omega \times \mathbb{T})$. Therefore, by the usual Fubini theorem, B has measure one, as we wanted to show.

We omit the second part of the proof, which is similar. ⊣

So far, it may appear to the reader that the adapted function and the conditional process approaches are the same in all respects. We have now come to a point where the two approaches differ. The Adapted Lifting Theorem above is not valid if we replace conditional processes by adapted functions! The reason is a little technical and has to do with the nature of liftings involving the filtration of the hyperfinite space. We will have a chance to look into this matter in Chapters 7 and 8.

2C. Universality theorem

We are now ready to prove two of the central results in the model theory of stochastic processes: the Adapted Universality and Adapted Homogeneity Theorems for hyperfinite adapted spaces. As we shall see in this chapter, these theorems explain to a large extent why hyperfinite adapted spaces have been a useful tool in stochastic analysis.

These results appeared in Keisler [1988] and Keisler [1986b]. Here we give their complete proofs, thanks to Javier Peña, who filled in all the details that were just sketched or suggested in the original papers. In particular, we give a full proof of the Adapted Universality Theorem that appeared in a mutilated form in Keisler [1988].

In Chapter 1 we introduced the notion of a universal probability space, which is a probability space that is so rich that it has all probability distributions. We saw in Theorem 1E.2 that the universal probability spaces are just the atomless probability spaces, so ordinary universal probability spaces are easily found. Hyperfinite probability spaces (without the filtration) are also universal. A simple direct proof using the Saturation Principle can be found in Keisler [1984] or Fajardo [1990b]. It is even simpler to show that hyperfinite probability spaces are atomless.

PROPOSITION 2C.1. *Every hyperfinite probability space Ω is atomless, and hence universal.*

Proof. Suppose A is a subset of Ω of positive Loeb measure. Then A has an internal subset B of positive Loeb measure. B has an internal subset C such that the internal cardinalities of C and $B \setminus C$ differ by at most one. Since Ω has the counting measure, it follows that the Loeb measure of C is half the Loeb measure of B. Thus C is a subset of A of positive Loeb measure that we are looking for.   ⊣

Let us now turn to the problem of finding a universal adapted space. We have promised to prove in this chapter that universal adapted spaces exist. We now keep this promise. We saw in Theorem 1E.11 that no ordinary adapted space is universal. Since the ordinary adapted spaces will not do the job, it is natural to look at the hyperfinite adapted spaces.

Theorem 2C.2. (*Adapted Universality Theorem*). *Every hyperfinite adapted space Ω is universal.*

Proof. By Theorem 1E.6, it suffices to prove that Ω is universal for random variables. Suppose that x is a random variable with values in M. Let C be a countable set of functions which is $\equiv$-dense, let $\mathbb{T}_n = \{\frac{1}{n}, \ldots, 1 - \frac{1}{n}, 1\}$ and let F_n be the set of all expressions of the form $f(t_1, \ldots, t_n)$, where $f(v_1, \ldots, v_n)$ is a conditional process built from functions in C and $t_1, \ldots, t_n \in \mathbb{T}_n$.

In order to have a better understanding of how the proof goes we have split it into four steps that capture the main ideas. Here they are.

Step I. For each $n \in \mathbb{N}$ there exists a family of internal random variables Y_f, $f \in F_n$, with the following properties:

(1) $(Y_f : f \in F_n)$ and $(f(x) : f \in F_n)$ have the same joint distributions, that is, the same distributions as random variables in $M^{\mathbb{N}}$.

(2) If $r < t$ in $\mathbb{T}_n$ and $f \in F_n$ then $Y_{E[f|s]} \approx E[Y_f|s]$ a.s.

Step II. For each $n \in \mathbb{N}$ there exists an internal random variable $Z_n : \Omega \to {}^*M$ such that for each $f \in F_n$, $f(Z_n) \approx Y_f$ a.s. In particular, by condition (1) of Step I, $E[f(Z_n)] \approx E[f(x)]$ for each $f \in F_n$.

Step III. By the Saturation Principle, there is an internal random variable $Y : \Omega \to {}^*M$ such that for each $n \in \mathbb{N}$ and $f \in F_n$, $E[f(Y)] \approx E[f(x)]$. Thus, $Y(\omega)$ has a standard part a.s., so we can define $y : \Omega \to M$ by setting $y = {}^oY$.

Step IV. The y defined in Step III satisfies $x \equiv y$.

We now give the details of each step.

Step I. For each $t \in \mathbb{T}_n$ with $t < 1$, let $F_n|t$ be the set of all $f \in F_n$ of the form $E[g|s]$ with $s \leq t$. Let $F_n|1 = F_n$. By induction on $t \in \mathbb{T}_n$, we construct a family of internal random variables Y_f, $f \in F_n$, with the following properties for each $t \in \mathbb{T}_n$.

(a) If $f \in F_n|t$ then Y_f depends only on $(\omega|t)$.

(b) The families $({}^oY_f : f \in F_n|t)$ and $(f(x) : f \in F_n|t)$ of random variables have the same joint distributions, that is, the same distributions as random variables in $M^{\mathbb{N}}$.

(c) If $r < t$ in $\mathbb{T}_n$ and $f \in F_n|t$ then $Y_{E[f|s]} \approx E[Y_f|s]$ a.s.

For the initial step $t = \frac{1}{n}$, notice that (c) is trivially satisfied. Given a finite $F \subseteq F_n|\frac{1}{n}$, by the universality theorem for hyperfinite probability spaces and random variables, there exists a family of internal random variables $Y_f : \Omega|\frac{1}{n} \to {}^*\mathbb{R}$, $f \in F$, such that $({}^oY_f : f \in F)$ and $(fx : f \in F)$ have the same distribution. By the Saturation Principle, we obtain a family of internal random variables $Y_f : \Omega|\frac{1}{n} \to {}^*\mathbb{R}$, $f \in F_n$ such that $({}^oY_f : f \in F_n)$ and $(f(x) : f \in F_n)$ have the same joint distribution.

The induction step from s to $t = s + \frac{1}{n}$ is as follows. We already have chosen Y_f for $f \in F_n|s$. For each $m \in \mathbb{N}$, partition $\mathbb{R}$ into half-open intervals of length $1/m$. For each $\omega \in \Omega$, $m \in \mathbb{N}$ and finite $G \subseteq F_n|s$, let $J(G, m, \omega)$ be the set of all $\lambda \in \Lambda$ such that for each $g \in G$, $g(x)(\lambda)$ is in the same $1/m$-interval as ${}^oY_g(\omega)$. $J(G, m, \omega)$ can be written as $\bigcap_{g \in G} (gx)^{-1}(I_g)$ where I_g is the $1/m$-interval such that ${}^oY_g(\omega) \in I_g$.

With this notation, let $J'(G, m, \omega) = \bigcap_{g \in G} {}^oY_g^{-1}(I_g)$. Note that $J(G, m, \omega)$ meets $J(G, m, \omega')$ if and only if $\omega' \in J'(G, m, \omega)$, and in that case $J(G, m, \omega) = J(G, m, \omega')$. Moreover, by condition (b) of the induction hypothesis,

$$Q[J(G, m, \omega)] = P[J'(G, m, \omega)].$$

So the sets $J(G, m, \omega)$ depend only on $(\omega|s)$ and form a partition of almost all of Λ.

Let H be a finite subset of $F_n|t$ and let $G = \{E[f|s] : f \in H\}$. When $Q[J(g, m, \omega)] > 0$, we may choose internal random variables $U_h^\omega, h \in H$, on the hyperfinite space $(\omega|s) = \{\omega' \in \Omega : \omega'|s = \omega|s\}$ such that the families $({}^oU_h^\omega : h \in H)$ and $(h(x) : h \in H)$ have the same joint distribution, and in addition $U_h^\omega(\omega')$ depends only on $\omega'|t$. There are only countably many distinct sets $J(G, m, \omega)$ of positive measure. Let $(J(G, m, \omega_i))_{i \in \mathbb{N}}$ be an enumeration of them. By our previous observations, $(J'(G, m, \omega_i))_{i \in \mathbb{N}}$ is a partition of almost all of Ω. Extend each $U_h^{\omega_i}$ over $J'(G, m, \omega_i)$ by letting $U_h^{\omega_i}(\omega) = U_h^{\omega_i}(\bar\omega)$ where $\bar\omega(h) = \omega_i(h)$ for $h \leq s$ and $\bar\omega(h) = \omega(h)$ for $h > s$. It is clear that the joint distribution of the family $({}^oU_h^{\omega_i} : h \in H)$ on $J'(G, m, \omega_i)$ is the same as the joint distribution of the family $({}^oU_h^{\omega_i} : h \in H)$ on $(\omega_i|s)$, which is also the same as the joint distribution of $(hx : h \in H)$ on $J(G, m, \omega_i)$. By the Saturation Principle, there exists a family of internal random variables $(U_h : h \in H)$ such that $U_h(\omega) = U_h^{\omega_i}(\omega)$ where $\omega \in J'(G, m, \omega_i)$. The joint distribution of $({}^oU_h : h \in H)$ on each $J'(G, m, \omega_i)$ is the same as the joint distribution of $(h(x) : h \in H)$ on $J(G, m, \omega)$, but $P[J'(G, m, \omega_i)] = Q[J(G, m, \omega_i)]$. So the families $({}^oU_h : h \in H)$ and $(h(x) : h \in H)$ have the same joint distribution on Ω and Λ, and $U_h(\omega)$ depends only on $(\omega|t)$.

Now suppose $\omega \in J'(G, m, \omega_i)$. Then $E[U_h|s](\omega) = E[U_h|s](\omega_i)$ by the definition of U_h. By the definition of internal conditional expectation $E[U_h|s]$ equals $E[U_h^{\omega_i}]$ on $(\omega_i|s)$ and by construction it is infinitely close to $E[h(x)]$ on $J(G, m, \omega_i)$. On the other hand, $E[h|s]$ on $J(G, m, \omega_i)$ equals $E[E[h(x)]]$ on $J(G, m, \omega_i)$, since $J(G, m, \omega_i)$ is $\sigma(E[h|s])$-measurable. Then it belongs to $I_{E[h|s]}$

by the definition of $J(G, m, \omega_i)$. So, for almost all $\omega \in \Omega$, $E[U_h|s](\omega)$ is within $1/m$ of $Y_{E[h|s]}(\omega)$.

We claim that by the Saturation Principle there is a family of internal random variables Y_f, $f \in F_n|t$, on Ω which satisfies (a) through (c)

To see this, apply the Saturation Principle to the following set of formulas, with v_f, $f \in F_n|t$, where k, m range over $\mathbb{N}$ and ϕ ranges over $C \cap C(\mathbb{R}^k, \mathbb{R})$:

$$v_f(\omega) \text{ depends only on } (\omega|t);$$

$$|E[\phi(v_{f_1}, \ldots v_{f_k})] - E[\phi(f_1 x, \ldots, f_k x)]| < \frac{1}{m};$$

$$P\left[|E[v_f|s] - Y_{E[f|s]}| \leq \frac{1}{m}\right] \geq 1 - \frac{1}{m}.$$

This set of formulas is finitely satisfiable because a finite subset of it determines a finite subset $H \subseteq F_n|t$ and an $m \in \mathbb{N}$, and then it is satisfiable by the previous discussion. When $t = 1$, we obtain the desired family.

Step II. Claim: The set of formulas

$$\left\{ P\left[|\phi(Z) - Y_\phi| \leq \frac{1}{m}\right] \geq 1 - \frac{1}{m} \right\},$$

where ϕ ranges over $C \cap C(M, \mathbb{R})$ and m ranges over $\mathbb{N}$, is finitely satisfiable.

To prove this claim let $\phi_1, \ldots, \phi_k \in C \cap C(M, \mathbb{R})$ and let $m \in \mathbb{N}$. The variables $({}^o Y_{\phi_1}, \ldots, {}^o Y_{\phi_k})$ and $(\phi_1 x, \ldots, \phi_k x)$ have the same joint distribution. The functions $\phi_1, \ldots, \phi_k$ are bounded, so there is a real M such that the variables $({}^o Y_{\phi_1}, \ldots, {}^o Y_{\phi_k})$ and $(\phi_1 x, \ldots, \phi_k x)$ take their values in $[-M, M]^k$. Partition $[-M, M]$ into intervals of length $1/2m$. Given a collection $I_1, \ldots, I_k$ of these intervals, we have

$$P[{}^o Y_{f_i} \in I_i, i = 1, \ldots, k] = Q[f_i x \in I_i, i = 1, \ldots, k].$$

Take an internal subset $\Omega_{I_1, \ldots, I_k}$ such that

$$P[\Omega_{I_1, \ldots, I_k} \triangle \{\omega \in \Omega : {}^o Y_{f_i}(\omega) \in I_i, i = 1, \ldots, k\}] = 0.$$

We can construct these sets so that they form a partition of Ω. Now take a point

$$y_{I_1, \ldots, I_k} \in \bigcap_{i=1}^{k} \phi_i^{-1}(I_i)$$

if this set is not empty, and take an arbitrarily fixed point in M if it is empty. Define $Z : \Omega \to {}^*M$ by

$$Z(\Omega_{I_1, \ldots, I_k}) = {}^*y_{I_1, \ldots, I_k}.$$

This random variable is internal because it is defined by a finite number of internal cases. By construction, for almost all ω the ${}^o Y_{\phi_i}(\omega)$ and ${}^o({}^*\phi_i(Z(\omega)))$ are in the same interval I_i, so the distance between $Y_{\phi_i}(\omega)$ and ${}^*\phi_i(Z(\omega))$ is less than $1/m$.

By the Saturation Principle, the set of formulas

$$P\left[|\phi(Z) - Y_\phi| \le \frac{1}{m}\right] \ge 1 - \frac{1}{m}$$

where ϕ ranges over $C \cap C(M, \mathbb{R})$ and m ranges over $\mathbb{N}$ is satisfied by an internal random variable Z_n. So we have $\phi(Z_n) \approx Y_\phi$ a.s. for $\phi \in C \cap C(M, \mathbb{R})$. Now we prove by induction that

(1) $$f(Z_n) \approx Y_f \text{ for all } f \in F_n.$$

Basis Step: By construction.
Composition Step: Suppose $f_1, \ldots, f_k \in F_n$ satisfy (1) and $\psi \in C \cap C(\mathbb{R}^k, \mathbb{R})$. We want to prove that

$$\psi(f_1, \ldots, f_k)(Z_n) \approx Y_{\psi(f_1, \ldots, f_k)} \, a.s.$$

By definition, we have

$$E[|\psi(f_1, \ldots, f_k)(x) - \psi(f_1(x), \ldots, f_k(x))|] = 0.$$

Now, by condition (b) of Step I,

$$\left({}^o Y_{\psi(f_1, \ldots, f_k)}, {}^o Y_{f_1}, \ldots, {}^o Y_{f_k}\right)$$

has the same joint distribution as

$$(\psi(f_1, \ldots, f_k)(x), f_1(x), \ldots, f_k(x)),$$

so

$$E[|Y_{\psi(f_1, \ldots, f_k)} - \psi({}^o Y_{f_1}, \ldots, {}^o Y_{f_k})|] = 0.$$

Then

$${}^o Y_{\psi(f_1, \ldots, f_k)} = \psi({}^o Y_{f_1}, \ldots, {}^o Y_{f_k}) \, a.s.$$

By the continuity of ψ we have

(2) $$Y_{\psi(f_1, \ldots, f_k)} \approx \psi(Y_{f_1}, \ldots, Y_{f_k}) \, a.s.$$

By continuity and induction hypothesis we have

(3) $$\psi(f_1, \ldots, f_k)(Z_n) = \psi(f_1(Z_n), \ldots, f_k(Z_n)) \approx \psi(Y_{f_1}, \ldots, Y_{f_k}) \, a.s.$$

Putting (2) and (3) together we obtain

$$\psi(f_1, \ldots, f_n)(Z_n) \approx Y_{\psi(f_1, \ldots, f_n)} \, a.s.$$

Conditional Expectation Step: Suppose $f \in F_n$ satisfies (1). By condition (b) of Step I, if $s \in \mathbb{T}_n$ then $E[Y_f|s] \approx Y_{E[f|s]}$ a.s. On the other hand, by induction hypothesis we have

$$E[f|s](Z_n) = E[f(Z_n)|s] \approx E[Y_f|s] \, a.s.$$

Therefore, $E[f|s]Z_n \approx Y_{E[f|s]} \, a.s.$

Step III. For each $n \in \mathbb{N}$ and each $f \in F_n$ let $C_{n,f}$ be the set of all internal $Y : \Omega \to {}^*M$ such that $|E[f(Y)] - E[f(x)]| < 1/n$. We claim that this family has the finite intersection property. To see this consider $(n_1, f_1), \cdots (n_k, f_k)$ with $f_i \in F_{n_i}$ for $i = 1, \ldots, k$. Let $N = n_1, \ldots, n_k$. Then $F_{n_i} \subseteq F_N$ and $1/N < 1/n_i$ for $i = 1, \ldots, k$. So $Z_N \in \bigcap_{i=1}^{k} C_{n_i,f_i}$ and the finite intersection property is verified. By the Saturation Principle, there is an internal function $Y : \Omega \to {}^*M$ such that for each $n \in \mathbb{N}$ and $f \in F_n$,

$$(4) \qquad\qquad E[f(Y)] \approx E[Y_f].$$

Now, to see that $Y(\omega)$ has a standard part a.s, let $\{a_i : i \in \mathbb{N}\}$ be a countable dense subset of M and define $f_n : M \to \mathbb{R}$ by $f_n(y) = d(y, \{a_1, \ldots, a_n\})$. Obviously, f_n is continuous, and for each $m \in \mathbb{N}$,

$$\bigcup_{n \in \mathbb{N}} \left\{ y \in M : f_n(y) < \frac{1}{m} \right\} = M.$$

Then

$$\bigcup_{n \in \mathbb{N}} \left\{ \lambda \in \Lambda : f_n \circ x(\lambda) < \frac{1}{m} \right\} = \Lambda,$$

so

$$\lim_{n \to \infty} Q \left[\left\{ \lambda \in \Lambda : f_n \circ x(\lambda) < \frac{1}{m} \right\} \right] \to 1.$$

Notice that

$$Q \left[\left\{ \lambda \in \Lambda : f_n \circ x(\lambda) < \frac{1}{m} \right\} \right] = E[I_m \circ f_n \circ x],$$

where $I_m : \mathbb{R} \to \mathbb{R}$ is the characteristic function of $[0, 1/m)$. The function $I_m \circ f_n$ can be approximated by functions built from C, so by condition (4) we have

$$E[I_m \circ f_n \circ Y] \approx E[I_m \circ f_n \circ x] = Q \left[\left\{ \lambda \in \Lambda : f_n \circ x(\lambda) < \frac{1}{m} \right\} \right].$$

Then

$$\lim_{n \to \infty} P \left[\left\{ \omega \in \Omega : f_n \circ Y(\omega) < \frac{1}{m} \right\} \right] = \lim_{n \to \infty} {}^o E[I_m \circ f_n \circ Y] \to 1.$$

So, for each $m \in \mathbb{N}$ we have

$$P \left[\bigcup_{n \in \mathbb{N}} \left\{ \omega \in \Omega : f_n \circ Y(\omega) < \frac{1}{m} \right\} \right] = 1.$$

Therefore

$$P \left[\bigcap_{m \in \mathbb{N}} \bigcup_{n \in \mathbb{N}} \left\{ \omega \in \Omega : f_n \circ Y(\omega) < \frac{1}{m} \right\} \right] = 1.$$

Given

$$\omega \in \bigcap_{m \in \mathbb{N}} \bigcup_{n \in \mathbb{N}} \left\{ \omega \in \Omega : f_n \circ Y(\omega) < \frac{1}{m} \right\},$$

we can obtain a sequence $(y_m)_{m \in \mathbb{N}}$ in M such that

$$*d(Y(\omega), *y_m) < \frac{1}{m}.$$

Since M is complete there is a $y \in M$ such that $d(*Y(\omega), *y) \approx 0$, so $Y(\omega)$ is nearstandard.

Step IV. Let g be a conditional process built from C. Y is a right lifting of $y = {}^o Y$, because they are random variables. So by the Adapted Lifting Theorem, $g(Y)$ is a right lifting of $g(y)$. On the other hand, x and y are both right continuous in probability because they are random variables, so $g(x)$ and $g(y)$ are right continuous in probability. In particular, given a real tuple $\vec{s}$ we have that if $\vec{u} \downarrow \vec{s}$ then $E[g(x)(\vec{u})] \to E[g(x)(\vec{s})]$. For each rational tuple $\vec{t}$, there is a $n \in \mathbb{N}$ such that $g(\vec{t}) \in F_n$. so

$$E[g(Y)(\cdot, \vec{t})] \approx E[g(x)(\cdot, \vec{t})].$$

We want to prove that

$$E[g(y)(\cdot, \vec{s})] = E[g(x)(\cdot, \vec{s})]$$

for each real tuple $\vec{s}$. Fix a real tuple $\vec{s}$. and let $\vec{u} \approx \vec{s}$ be such that if $\vec{r} \geq \vec{u}$ and $\vec{r} \approx \vec{s}$, then $g(Y)(\cdot, \vec{r})$ is a lifting of $g(y)(\cdot. \vec{s})$. We claim that there exists $\vec{w} \geq \vec{u}$, $\vec{w} \approx \vec{s}$ such that

$$E[g(Y)(\cdot, \vec{w})] \approx E[g(x)(\cdot, \vec{s})].$$

Otherwise for each $n \in {}^* \mathbb{N} - \mathbb{N}$ we have

$$|E[g(Y)(\cdot, \vec{w})] - E[g(x)(\cdot. \vec{s})]| > \frac{1}{n}$$

for all $\vec{w}$ with $\vec{u} \leq \vec{w} \leq \vec{u} + 1/n$, and then by underflow there is $n \in \mathbb{N}$ with this property. Let $\vec{t}$ be a rational tuple such that $\vec{u} \leq \vec{t} \leq \vec{u} + 1/n$ and

$$|E[g(x)(\vec{s})] - E[g(x)(\vec{t})]| < \frac{1}{2n}$$

(such a $\vec{t}$ exists because $g(x)$ is right continuous in probability). Then

$$|E[g(Y)(\vec{t})] - E[g(x)(\vec{t})]| >$$

$$|E[g(Y)(\vec{t})] - E[g(x)(\vec{s})]| - |E[g(x)(\vec{s})] - E[g(x)(\vec{t})]| > \frac{1}{n} - \frac{1}{2n} = \frac{1}{2n},$$

which is a contradiction! This proves the claim that $\vec{w}$ exists. Therefore, we have

$$E[g(y)(\vec{s})] = {}^o E[g(Y)(\vec{w})] = E[g(x)(\vec{s})]$$

for each conditional processes g built from C and for each real tuple $\vec{s}$. Thus $x \equiv y$. $\dashv$

In a κ-saturated nonstandard universe, one can get a stronger form of the universality theorem by modifying the above proof to handle a whole family of stochastic processes at once.

THEOREM 2C.3. *In a κ-saturated nonstandard universe, every hyperfinite adapted space Ω is κ-universal, in the sense that for every family of stochastic process $(x_\alpha)_{\alpha<\kappa}$ on another adapted space Λ there is a family of stochastic process $(y_\alpha)_{\alpha<\kappa}$ on Ω such that*

$$(\Lambda, (x_\alpha)_{\alpha<\kappa}) \equiv (\Omega, (y_\alpha)_{\alpha<\kappa}).$$ ⊣

At this point it is worth mentioning that in the literature there are extensions of the notions of stochastic processes and adapted spaces which have two or more time parameters. There has been work leading toward a nonstandard theory of processes with two time parameters (see for example Dalang [1989]), and a corresponding probability logic (see Fajardo [1985b]). Peña [1993] proved a universality theorem for the two-parameter hyperfinite adapted spaces introduced by Dalang [1989].

2D. Homogeneity theorem

The homogeneity property which is essential in the nonstandard approach to stochastic analysis (see Keisler [1984]) is rarely mentioned in the literature. Let us take a look the meaning of homogeneity for plain probability spaces before taking up the adapted case. By an **almost sure bijection** from a probability space Ω to a probability space Γ we mean a bijection from a subset of Ω of measure one to a subset of Γ of measure one.

DEFINITION 2D.1. *A probability space $(\Omega, \mathcal{F}, P)$ is said to be* **homogeneous** *if given two random variables x and y on Ω with the same distribution (i.e., $x \equiv_0 y$), there is an a.s. bijection h from Ω to Ω which preserves $\mathcal{F}$-measurability and P-measures, such that $x(\omega) = y(h(\omega))$ a.s.*

In a homogeneous probability space the usual intuition that "two random variables with the same distribution are the same" is correct, and is confirmed by the existence of a probabilistic automorphism.

Finite probability spaces with uniform counting measures are obviously homogeneous. We will see in the next chapter that no atomless ordinary probability space is homogeneous. Thus, for example, the unit interval with Lebesgue measure is not homogeneous, and the space of continuous paths with the Wiener measure is not homogeneous. The following result gives us some examples of atomless homogeneous probability spaces.

PROPOSITION 2D.2. *Every hyperfinite probability space Ω is homogeneous.*

PROOF. Let x and y be two random variables on Ω with values in a Polish space M which have the same law. We shall find an internal automorphism sending x to y. Let (S_n) be a countable open basis for M. By induction on n we can construct two sequences of internal sets $(C_n), (D_n)$ such that

$$P[C_n \Delta (x^{-1}(S_n))] = 0 = P[D_n \Delta (y^{-1}(S_n))].$$

Each Boolean combination of $C_1, \ldots, C_n$ has the same internal cardinality as the corresponding Boolean combination of $D_1, \ldots, D_n$.

Then for each n there is an internal bijection h_n of Ω which maps C_m onto D_m for each $m \leq n$. By the Saturation Principle there is an internal bijection h which maps C_n onto D_n for each $n \in \mathbb{N}$. h sends any Loeb measurable set to a Loeb measurable set with the same measure. Moreover, $y(h(\omega)) = x(\omega)$ a.s., as required.   ⊣

The preceding result will be used as the basis step of an induction in the proof of a deeper result, the Adapted Homogeneity Theorem, later in this section.

We now introduce the analog for adapted spaces of the model-theoretic notion of an isomorphism between two structures.

DEFINITION 2D.3.   (a) Let Ω, Γ be adapted spaces.

An **adapted isomorphism** from Ω to Γ is an a.s. bijection from Ω to Γ such that for each $t \in [0, 1]$. h maps $\mathcal{F}_t$ onto $\mathcal{G}_t$ (up to null sets) and preserves measures. An adapted isomorphism from Ω to Ω is called an **adapted automorphism**.

(b) Two processes x on Ω and y on Γ are **isomorphic**, in symbols $x \simeq y$, if there is an adapted isomorphism h from Ω to Γ such that for each $t \in [0, 1]$, $x(\omega, t) = y(h(\omega), t)$ a.s. x and y are **automorphic** if they are isomorphic and are defined on the same adapted space Ω.

(c) The adapted space Ω is said to be **homogeneous** if for every pair of stochastic processes x and y defined on Ω, $x \equiv y$ implies $x \simeq y$.

(d) If the condition in (c) holds when x. y are random variables, Ω is **homogeneous for random variables**.

The notion of an adapted isomorphism can be thought of as the strongest possible relation which preserves all probabilistic properties of stochastic processes. The reason for taking a.s. bijections instead of ordinary bijections is to leave room for a change in the mapping on a set of measure zero. For example. the relation of being isomorphic cannot be destroyed simply be adding a null set of large cardinality to a probability space.

Here is an example of an ordinary adapted space which is homogeneous and has an atomless probability measure. We leave the proof as an exercise for the reader.

EXAMPLE 2D.4.   Let $(\Omega_0. \mathcal{B}_0. P_0)$ be a finite set with the counting probability measure. Form the countable product space $(\Omega. \mathcal{B}. P) = (\Omega_0. \mathcal{B}_0. P_0)^{\mathbb{N}}$. Take an increasing sequence t_n of times converging to 1. Let $\mathcal{F}_t$ be the filtration such that for each $t \in [0, 1]$. ω and ω' belong to the same sets in $\mathcal{F}_t$ if and only if $\omega(n) = \omega'(n)$ whenever $t_n \leq t$. Then $(\Omega. \mathcal{F}_t. P)$ is a homogeneous adapted space with an atomless probability measure.

The adapted space in the above example falls far short of being universal. For instance, it does not support a Brownian motion, and an adapted process can take only finitely many values at times less than 1. Although the above example is atomless as a probability space, it is not atomless as an adapted space in a sense that will be introduced in Chapter 7.

Our next theorem shows that the hyperfinite adapted spaces are homogeneous as well as universal. The proof was originally sketched in Keisler [1986b] and Keisler [1988].

LEMMA 2D.5. *An adapted space is homogeneous if and only if it is homogeneous for random variables.*

PROOF. We prove the hard direction. Suppose Ω is homogeneous for random variables, and let x, y be stochastic processes with values in M on Ω such that $x \equiv y$. By Lemma 1E.3, there is a Borel function φ and cadlag processes z, w on Ω such that $z \equiv w$, $x = \varphi(z)$ a.s., and $y = \varphi(w)$ a.s. We can take z and w to be random variables in the Polish space $D([0, 1], M)$. By hypothesis there is an adapted automorphism h of Ω such that $z(\omega) = w(h(\omega))$ a.s. Then for each t we have $x(\omega, t) = y(h(\omega, t))$ a.s. $\dashv$

THEOREM 2D.6. (*Adapted Homogeneity Theorem*). *Every hyperfinite adapted space Ω is homogeneous. Moreover the adapted automorphism can be chosen to be an internal bijection from Ω to Ω.*

PROOF. In view of Lemma 2D.5, it suffices to prove that Ω is homogeneous for random variables. Suppose x, y are random variables on Ω such that $x \equiv y$. Let C be a countable $\equiv$-dense set, and let X and Y be liftings of x and y. By the Adapted Lifting Theorem, for each conditional expression f, $f(X)$ and $f(Y)$ are right liftings of $f(x)$ and $f(y)$. As we did for the Adapted Universality Theorem, we break the proof into steps to make it easier to understand the main ideas. Let $n \in \mathbb{N}$ and

$$\mathbb{T}_n = \left\{ \frac{1}{n}, \dots, 1 - \frac{1}{n}, 1 \right\}.$$

Step I.
There exist $b_1, \dots, b_n \in \mathbb{T}$ such that ${}^\circ b_i = i/n$ for $i = 1, \dots, n$, and for every conditional process g built from C and every $\vec{b}$ in $\{b_1, \dots, b_n\}$,

$$E[g(X)(\cdot, \vec{b})] \approx E[g(Y)(\cdot, \vec{b})].$$

Step II.
Take $b_1, \dots, b_n$ as in Step I. There is an internal bijection $h_n : \Omega \to \Omega$ such that h_n maps $(\omega|b_k)$ onto $(h_n(\omega)|b_k)$ for each ω and $k = 1, \dots, n$, and

$$E[g(X)(\cdot, \vec{b})] \approx E[g(Y)(h_n(\cdot), \vec{b})]$$

for all conditional process built g from C and all $\vec{b}$ in $\{b_1, \dots, b_n\}$. In particular, $X(\omega) \approx Y(h_n(\omega))$ a.s. This bijection is not adapted but "looks ahead" to the next b_k. That is, if $b_k \leq \underline{t} \leq b_{k+1}$ and $\underline{t} \in \mathbb{T}$, then

$$\mathcal{A}_{b_k} \subseteq h_n(\mathcal{A}_{\underline{t}}) \subseteq \mathcal{A}_{b_{k+1}}.$$

Step III.
By overspill, we get a hyperfinite K and an internal bijection $h_K : \Omega \to \Omega$ such that $X(\omega) \approx Y(h_K(\omega))$ a.s. and for all $\underline{t} \in \mathbb{T}$,

$$\mathcal{A}_{\underline{t}-1/K} \subseteq h_K(\mathcal{A}) \subseteq \mathcal{A}_{\underline{t}+1/K}.$$

This h_K is an internal adapted automorphism of Ω and sends x to y.

Let us see the details of each step:

Step I.

Let $g_1, \dots, g_m$ be conditional processes built from C and $\vec{t} = (1/k_1, \dots, 1/k_m)$ in $\mathbb{T}_n$. $g_j(X)$ and $g_j(Y)$ are right liftings of $g_j(x)$ and $g_j(y)$ for $j = 1, \dots, m$. Hence there are $b_1, \dots, b_n \in \mathbb{T}$ with ${}^o b_i = i/n$ such that $\vec{b} \approx \vec{t}$ and whenever $\vec{c} \geq \vec{b}$ and $\vec{c} \approx \vec{t}$, then $g_j(X)(\cdot, \vec{c})$ and $g_j(Y)(\cdot, \vec{c})$ are liftings of $g_j(x)(\cdot, \vec{t})$ and $g_j(y)(\cdot, \vec{t})$. Therefore

$$E[g_j(X)(\cdot, \vec{b})] \approx E[g_j(x)(\cdot, \vec{t})] = E[g_j(y)(\cdot, \vec{t})] \approx E[g_j(Y)(\cdot, \vec{b})].$$

By using the fact that a countable set of infinitesimals has an infinitesimal upper bound, we can obtain $b_1, \dots, b_n \in \mathbb{T}$ with ${}^o b_i = i/n$ such that for each conditional process g built from C, we have $E[g(X)(\cdot, \vec{b})] \approx E[g(Y)(\cdot, \vec{b})]$ for all $\vec{b}$ in $\{b_1, \dots, b_n\}$.

Step II.

Let us write $E[Y, \Gamma]$ for the expectation of a random variable Y on a subset Γ of Ω. By induction on $i = 1, \dots, n$. we will construct internal bijections $h^i : \Omega|b_i \to \Omega|b_i$ such that

$$h^{i-1}(\omega|b_{i-1}) = (h^i(\omega|b_i)|b_{i-1}) \text{ for } i > 1,$$

and such that for all g built from C and all $\vec{b}$ in $\{b_1, \dots, b_n\}$,

$$E[g(X)(\cdot, \vec{b}), (\omega|b_i)] \approx E[g(Y)(\cdot, \vec{b}), h^i(\omega|b_i)]$$

for almost all $(\omega|b_i) \in \Omega|b_i$. The induction step from i to $i + 1$ is as follows.

Assume that we have an h^i with the above property. We claim that for almost all $\omega \in \Omega$ there exists a bijection

$$h_\omega : (\omega|b_i)|b_{i+1} \to h^i(\omega|b_i)|b_{i+1}$$

(that is, between the parts of $\Omega|b_{i+1}$ which refine $(\omega|b_i)$ and $h^i(\omega|b_i)$) such that for all g built from C and all $\vec{b}$ in $\{b_1, \dots, b_n\}$,

$$E[g(X)(\cdot, \vec{b}), (\omega'|b_{i+1})] \approx E[g(Y)(\cdot, \vec{b}), h_\omega(\omega'|b_{i+1})]$$

for almost all

$$(\omega'|b_{i+1}) \in (\omega|b_i)|b_{i+1}.$$

To see this, note that given $g_1, \dots, g_k, \vec{b}_1, \dots, \vec{b}_k$, and $\psi \in C \cap C(\mathbb{R}^k, \mathbb{R})$, by the induction hypothesis, for almost all $(\omega|b_i) \in \Omega|b_i$ we have:

$$E[\psi(g_1(X)(\cdot, \vec{b}_1), \dots, g_k(X)(\cdot, \vec{b}_k)), (\omega|b_i)] \approx$$

$$E[\psi(g_1(Y)(\cdot, \vec{b}_1), \dots, g_k(Y)(\cdot, \vec{b}_k)), h^i(\omega|b_i)].$$

Then the random variables $({}^o X_{g_j, \vec{b}_j} : j = 1, \dots, k)$ and $({}^o Y_{g_j, \vec{b}_j} : j = 1, \dots, k)$ have the same finite dimensional distribution on $(\omega|b_i)|b_{i+1}$ and $h^i(\omega|b_i)|b_{i+1}$

respectively. Therefore by Proposition 2D.2, the homogeneity property for hyperfinite probability spaces, there exists an internal bijection

$$h : (\omega|b_i)|b_{i+1} \to h^i(\omega|b_i)|b_{i+1}$$

such that for almost all $(\omega'|b_{i+1}) \in (\omega|b_i)|b_{i+1}$,

$$E[g_j(X)(\cdot,\vec{b}_j), (\omega'|b_{i+1})] \approx E[g_j(Y)(\cdot,\vec{b}_j), h(\omega'|b_{i+1})].$$

By the Saturation Principle, we obtain h_ω for almost all $\omega \in \Omega$ with the required properties required. This proves the Claim.

Now apply transfer to the following sentence:

For all finite sets Ω, all finite $\mathbb{T} \subseteq \mathbb{R}$, all $b_1, b_2 \in \mathbb{T}$ with $b_1 < b_2$, all $X, Y : \Omega^{\mathbb{T}} \to \mathbb{R}$, all conditional processes $g_1, \ldots, g_k$ built from C, and all $m \in \mathbb{N}$,

if

$h : \Omega^{\mathbb{T}}|b_1 \to \Omega^{\mathbb{T}}|b_1$ is a bijection such that

$$P\left[\left\{(\omega|b_1) \in \Omega^{\mathbb{T}}|b_1 : |E[g_j(X),(\omega|b_1)] - E[g_j(Y), h(\omega|b_1)]| < \frac{1}{m}\right\}\right] \geq 1 - \frac{1}{m}$$

for each $j \leq k$, and the set of $(\omega|b_1) \in \Omega^{\mathbb{T}}$ such that there is a bijection h_ω on $(\omega|b_1) \cap \Omega^{\mathbb{T}}|b_2$ with

$$P\left[\left\{(\omega|b_2) \in (\omega|b_1) \cap \Omega^{\mathbb{T}}|b_2 \right.\right.$$

$$\left.\left. : |E[g_j(X),(\omega|b_2)] - E[g_j(Y), h_\omega(\omega|b_2)]| < \frac{1}{m}\right\}\right] \geq 1 - \frac{1}{m}$$

for each $j \leq k$ has probability at least $1 - \frac{1}{m}$,

then

there is a bijection $g : \Omega^{\mathbb{T}}|b_2 \to \Omega^{\mathbb{T}}|b_2$ such that $g(\omega|b_1) = h(\omega|b_1)$ for all $(\omega|b_1) \in \Omega^{\mathbb{T}}|b_1$, and for each $j \leq k$,

$$P\left[\left\{(\omega|b_1) \in \Omega^{\mathbb{T}}|b_1 : |E[g_j(X),(\omega|b_1)] - E[g_j(Y), h(\omega|b_1)]| < \frac{1}{m}\right\}\right] \geq 1 - \frac{1}{m}.$$

Given conditional processes $g_1, \ldots, g_k$ built from $C, \vec{b}_1, \ldots, \vec{b}_k$ in $\{b_1, \ldots, b_n\}$, and $m \in \mathbb{N}$, there is an internal bijection $h : \Omega^{\mathbb{T}}|b_{i+1} \to \Omega^{\mathbb{T}}|b_{i+1}$ such that $h(\omega'|b_i) = h^i(\omega'|b_i)$ for all $\omega' \in \Omega^{\mathbb{T}}|b_{i+1}$, and for each $j \leq k$, the set of all $(\omega|b_{i+1}) \in \Omega^{\mathbb{T}}|b_{i+1}$ such that

$$|E[g_j(X)(\cdot,\vec{b}_j), (\omega'|b_{i+1})] - E[g_j(Y)(\cdot,\vec{b}_j), h(\omega'|b_{i+1})]| < \frac{1}{m}$$

has probability at least $1 - \frac{1}{m}$.

By the Saturation Principle, we obtain an internal bijection

$$h^{i+1} : \Omega^{\mathbb{T}}|b_{i+1} \to \Omega^{\mathbb{T}}|b_{i+1}$$

with the required property that $h^i(\omega'|b_i) = (h^{i+1}(\omega'|b_{i+1})|b_i)$ and for each $\omega' \in \Omega^{\mathbb{T}}|b_{i+1}$, each conditional process g built from C, and each $\vec{b}$ in $\{b_1, \dots, b_n\}$,

$$E[g_j(X)(\cdot, \vec{b}_j), (\omega'|b_{i+1})] \approx E[g_j(Y)(\cdot, \vec{b}_j), h^{i+1}(\omega'|b_{i+1})]$$

for almost all $(\omega|b_{i+1}) \in \Omega^{\mathbb{T}}|b_{i+1}$.

The initial step of the induction can be seen as a particular case of this argument by taking $b_0 = 0$, so that $\Omega^{\mathbb{T}}|b_0 = \{\Omega^{\mathbb{T}}\}$, and taking

$$h^0 : \Omega^{\mathbb{T}}|b_0 \to \Omega^{\mathbb{T}}|b_0$$

as the only possible bijection.

The last induction step gives us a bijection h^n of Ω. We let $h_n = h^n$. By construction it is clear that

$$h_n(\omega|b_k) = h_n(\omega)|b_k \text{ for } k = 1, \dots, n,$$

and

$$E[g(X)(\cdot, \vec{b})] \approx E[g(Y)(h_n(\cdot), \vec{b})] \text{ for all } \vec{b} \text{ in } b_1, \dots, b_n.$$

In particular, $X(\omega) \approx Y(h_n(\omega))$ a.s.

It follows that $h_n(A_{b_k}) = A_{b_k}$. Therefore if $b_k \leq t \leq b_{k+1}$ then $A_{b_k} \subseteq h_n(A_t) \subseteq A_{b_{k+1}}$, that is, h_n looks ahead $1/n$.

Step III.

By Step II, for each $n \in \mathbb{N}$ there is an internal bijection $h_n : \Omega \to \Omega$ such that

$$P\left[|X(\omega) - Y(h_n(\omega))| < \frac{1}{n}\right] \geq 1 - \frac{1}{n}$$

and for all $t \in \mathbb{T}$,

$$A_{t-1/n} \subseteq h_n(A_t) \subseteq A_{t+1/n}.$$

By overspill, we obtain a $K \in {}^*\mathbb{N} - \mathbb{N}$ and an internal bijection $h_K : \Omega \to \Omega$ such that $X(\omega) \approx Y(h_K(\omega))$ a.s. and for all $\underline{t} \in \mathbb{T}$, $A_{\underline{t}-1/K} \subseteq h_K(A_{\underline{t}}) \subseteq A_{\underline{t}+1/K}$. Now, we know that $\mathcal{F}_t = \bigcap_{°\underline{t}>t} \sigma(A_{\underline{t}}) \vee \mathcal{N}$, so

$$h_K(\mathcal{F}_t) = \bigcap_{°\underline{t}>t} \sigma(h_K(A_{\underline{t}})) \vee \mathcal{N}.$$

Then

$$\bigcap_{°\underline{t}>t} \sigma(A_{\underline{t}-1/K}) \vee \mathcal{N} \subseteq h_K(\mathcal{F}_t) \subseteq \bigcap_{°\underline{t}>t} \sigma(A_{\underline{t}+1/K}) \vee \mathcal{N}.$$

But since K is hyperfinite,

$$\bigcap_{°\underline{t}>t} \sigma(A_{\underline{t}-1/K}) \vee \mathcal{N} = \bigcap_{°\underline{t}>t} \sigma(A_{\underline{t}}) \vee \mathcal{N} = \bigcap_{°\underline{t}>t} \sigma(A_{\underline{t}+1/K}) \vee \mathcal{N} = \mathcal{F}_t.$$

Thus h_K is an internal adapted automorphism of Ω, and

$$x(\omega) = {}^°X(\omega) = {}^°Y(h_K(\omega)) = y(h_K(\omega)) \text{ a.s.}$$

For the case that x and y are stochastic processes, we use Lemma 1E.3 to obtain a sequence $(t_n)_{n \in \mathbb{N}}$ of elements in $[0, 1]$ and two Borel functions $\psi : [0, 1] \to 2^{\mathbb{N}}$, $\phi : 2^{\mathbb{N}} \to M$ such that for all $t \in [0, 1]$,

$$x(\omega, t) = \phi(\psi(t, \langle x(\omega, t_n) : n \in \mathbb{N} \rangle)) \ a.s.$$

Since $x \equiv y$,

$$y(\omega, t) = \phi(\psi(t, \langle y(\omega, t_n) : n \in \mathbb{N} \rangle)) \ a.s.$$

and

$$\langle x(\omega, t_n) : n \in \mathbb{N} \rangle \equiv \langle y(\omega, t_n) : n \in \mathbb{N} \rangle.$$

By applying the result that we have just proved for random variables we obtain an internal adapted isomorphism $h : \Omega \to \Omega$ such that

$$\langle x(\omega, t_n) : n \in \mathbb{N} \rangle = \langle y(h(\omega), t_n) : n \in \mathbb{N} \rangle \ a.s.$$

Thus for each $t \in [0, 1]$,

$$x(\omega, t) = \phi(\psi(t, \langle x(\omega, t_n) : n \in \mathbb{N} \rangle)) =$$

$$\phi(\psi(t, \langle y(h(\omega), t_n) : n \in \mathbb{N} \rangle)) = y(h(\omega), t) \ a.s.$$

Then h is an internal adapted isomorphism which sends y to x.   ⊣

2E. Applications

The Universality and Homogeneity Theorems have several interesting consequences. We first mention a consequence of the Homogeneity Theorem that supports the claim that two process with the same adapted distribution have the same probabilistic properties. In a hyperfinite adapted space, two processes with the same adapted distribution are automorphic. This is definitely a good sign.

We can give two examples that have been powerful tools for applications in stochastic analysis (see Keisler [1984]). These examples illustrate perfectly why and how things are easier on hyperfinite adapted spaces. The first example concerns Anderson's construction of Brownian motion, which can be found in Anderson [1976] or in any of the basic nonstandard probability references given in this book.

PROPOSITION 2E.1. *On a hyperfinite adapted space Ω, every Brownian motion is automorphic to Anderson's Brownian motion.*

PROOF. Recall that every Brownian motion is a Markov process, and all Brownian motions have the same finite dimensional distribution. Then by Proposition 1D.3, any two Brownian motions have the same adapted distribution. Therefore, by the Adapted Homogeneity Theorem, any Brownian motion on Ω is automorphic to Anderson's Brownian motion.   ⊣

The main consequence of this proposition is, again intuitively, that anything that involves some Brownian motion in stochastic analysis can be reduced to Anderson's Brownian motion. Anderson's construction has two key features: it is very simple and easy to use, and at the same time it captures the physical intuition that a Brownian motion is a limit of random walks. The books Albeverio, Fenstad, Hoegh-Krohn, and Lindstrøm [1986] and Stroyan and Bayod [1986] and the memoir Keisler [1984] have plenty of examples backing up this statement.

The following result, taken from Keisler [1988], is yet another illustration of how the model theoretic approach that we are describing in these notes can be used within stochastic analysis. It deals with Itô's famous formula which is known as the fundamental theorem of stochastic calculus.

THEOREM 2E.2. (*Itô's formula*). *Let* Γ *be an adapted space,* $f : \Gamma \times [0, 1] \to \mathbb{R}$ *be bounded and adapted,* W *be a Brownian motion on* Γ,

$$y(\gamma, t) = \int_0^t f(\gamma, s) dW(\gamma, s),$$

and

$$\phi \in C^2(\mathbb{R}, \mathbb{R}).$$

Then

$$\phi(y(\gamma, t)) =$$

$$\phi(y(\gamma, 0)) + \int_0^t \phi'(y(\gamma, s)) f(\gamma, s) dW(\gamma, s) + \frac{1}{2} \int_0^t \phi''(y(\gamma, s))(f(\gamma, s))^2 ds.$$

PROOF. Anderson [1976] proved that Itô's formula is true for Anderson's Brownian motion. Now suppose that the formula has a counterexample in some adapted space. Since hyperfinite adapted spaces are universal, it has a counterexample for a Brownian motion on a hyperfinite adapted space Ω. Thus by Proposition 2E.1, it has a counterexample with respect to Anderson's Brownian motion, contrary to Anderson's result.   ⊣

We are now ready to take a look at a notion of a saturated adapted space. This notion originated with the study of hyperfinite adapted spaces, and is a central tool in the use of model theoretic ideas in applications of nonstandard analysis.

CHAPTER 3

SATURATED SPACES

In the preceding chapter we proved that hyperfinite adapted spaces have two key properties: universality and homogeneity. The paper Hoover and Keisler [1984] introduced and studied the weaker notion of a saturated adapted space. The saturated adapted spaces include the hyperfinite adapted spaces and the more general class of Loeb adapted spaces built using nonstandard analysis (see the discussion following Definition 2A.2). Another construction of saturated adapted spaces which is more difficult but avoids nonstandard analysis will be given in Chapter 6.

Saturated adapted spaces are rich enough to contain solutions of many existence problems. Our purpose in this chapter is to study the saturated adapted spaces and explore some their uses.

We will now be using the word "saturation" in two very different contexts—the Saturation Principle in the underlying nonstandard universe, and the notion of a saturated adapted space. Both of these are rooted in the notion of a saturated structure in first order model theory. Roughly speaking, the Saturation Principle says that the nonstandard universe is "saturated for bounded quantifier first order formulas," while a saturated adapted space is "saturated for adapted distributions." We hasten to add that the reader will not need to know anything about saturated structures in model theory, since our treatment will not depend on them.

3A. The saturation theorem

We begin by defining saturated probability and adapted spaces, and proving an easy theorem giving the relationship between the notions of saturation, universality, and homogeneity.

DEFINITION 3A.1. (a) *An adapted space Ω is said to be* **saturated** *if for every pair of stochastic processes x, y on an arbitrary adapted space Λ and every stochastic process x' on Ω such that $x' \equiv x$, there exists a stochastic process y' on Ω such that $(x', y') \equiv (x, y)$.*

(b) *Ω has the* **back and forth property** *if for all stochastic processes x, x', and y on Ω such that $x \equiv x'$, there exists y' on Ω such that $(x, y) \equiv (x', y')$.*

43

The following lemma is proved in the same way as Lemma 1E.6, using Lemma 1E.3. The proof is left as an exercise.

LEMMA 3A.2. *An adapted space has the back and forth property if and only if it has the back and forth property for random variables.*   ⊣

By replacing the words "stochastic process" everywhere by "random variable" and ≡ everywhere by ≡₀, we get the corresponding notions of a saturated probability space and the back and forth property for a probability space.

The notion of a saturated adapted space captures the intuitive idea of an adapted space which satisfies question (2) in the Introduction. Given a set S of adapted distributions of pairs of processes (x, y), and a process x on Ω, one can consider the following problem $\mathbb{P}(x, S)$: *Does there exist a process y on Ω such that the adapted distribution of (x, y) belongs to S?* By a **weak solution** of the problem we mean a pair of processes (x', y') on some adapted space such that $x' \equiv x$ and (x', y') has adapted distribution in S. Thus an adapted space Ω is saturated if and only if every problem of the form $\mathbb{P}(x, S)$ which has a weak solution has a solution y on Ω.

THEOREM 3A.3. (*Saturation Theorem*) *Let Ω be either a probability space or an adapted space.*

(*a*) Ω *is saturated if and only if it is universal and has the back and forth property.*

(*b*) *If Ω is homogeneous then it has the back and forth property.*

(*c*) *If Ω is universal and homogeneous then it is saturated.*

PROOF. (a) It is clear that saturation implies universality and the back and forth property. Assume that Ω is universal and has the back and forth property. Consider two stochastic processes x and y on an arbitrary adapted space Λ and a stochastic process x' on Ω such that $x \equiv x'$. By the universality of Ω we can find a pair (x'', y'') on Ω such that $(x'', y'') \equiv (x, y)$. Then we have $x' \equiv x''$. and by the back and forth property, there exists a stochastic process y' on Ω such that

$$(x', y') \equiv (x'', y'') \equiv (x, y).$$

This shows that Ω is saturated.

(b) Suppose Ω is homogeneous. Let $x. x'$. and y be processes on Ω such that $x \equiv x'$. Then there is an adapted automorphism h of Ω such that for all t, we have $x'(\omega, t) = x(h(\omega), t) a.s.$ Define $y'(\omega. t)$ as $y(h(\omega), t)$. Then $(x', y') \equiv (x, y)$, so Ω has the back and forth property.

(c) is an immediate consequence of (a) and (b).   ⊣

We now make sure that saturated probability and adapted spaces exist.

COROLLARY 3A.4. (*a*) *Every hyperfinite probability space is saturated.*

(*b*) *Every hyperfinite adapted space is saturated.*

PROOF. (a) By Proposition 2D.2 and the Saturation Theorem.

(b) By the Adapted Universality and Homogeneity Theorems and the Saturation Theorem.   ⊣

In the next two sections we will study saturated probability spaces and saturated adapted spaces respectively.

3B. Saturated and homogeneous probability spaces

Now that we know that saturated probability spaces exist, it is natural to ask which probability spaces are saturated. In this section we give some answers to this question. We will also consider the question of which probability spaces are homogeneous.

As we have seen, it is easy to find ordinary universal probability spaces. Are there any ordinary saturated probability spaces? The next result (from Hoover and Keisler [1984]) shows that the answer is "NO".

THEOREM 3B.1. *No ordinary probability space is saturated.*

PROOF. Let Ω be an ordinary probability space, and suppose that Ω is saturated. Let $C_0, C_1, \ldots$ be a countable open basis for the topology of Ω. Let x be the random variable on Ω with values in $2^{\mathbb{N}}$ such that for each $n \in \mathbb{N}$, $x(\cdot)(n)$ is the characteristic function of C_n. Consider the product space $\Omega \times 2$ where $2 = \{0, 1\}$ has the uniform measure. Let x', y' be the random variables on $\Omega \times 2$ such that $x'(\omega, t) = x(\omega), y'(\omega, t) = t$. Then $x \equiv_0 x'$, so there is a 2-valued random variable y on Ω such that $(x, y) \equiv_0 (x', y')$. Then y is the characteristic function of a Borel set $D \subseteq \Omega$ of measure $1/2$ which is independent of each of the sets C_n, contradicting the assumption that the sets C_n form an open basis for Ω. ⊣

COROLLARY 3B.2. *No ordinary atomless probability space is homogeneous.* ⊣

Although no ordinary probability space is saturated, every atomless probability space has the following weak saturation property.

PROPOSITION 3B.3. *Let Ω be an atomless probability space, and let Γ be another probability space. Let M and N be Polish spaces. Then for every simple M-valued random variable x on Ω and every pair of random variables $\bar{x}, \bar{y}$ on Γ with values in M and N such that $\bar{x} \equiv_0 x$, there exists a random variable y on Ω such that $(x, y) \equiv_0 (\bar{x}, \bar{y})$.*

PROOF. Let $\{m_1, \ldots m_k\}$ be the range of x, and let $A_j = x^{-1}(\{m_j\})$ and $B_j = \bar{x}^{-1}(\{m_j\})$. We may assume without loss of generality that each A_j has positive measure. Let $\bar{y}_j$ be the restriction of $\bar{y}$ to the set B_j. Since the restriction of Ω to A_j is atomless and B_j has the same measure as A_j, there is a random variable y_j on A_j such that $y(\omega) = y_j(\omega)$ whenever $\omega \in A_j$ for $j = 1, \ldots, n$. Then $(x, y) \equiv_0 (\bar{x}, \bar{y})$. ⊣

Note that the converse of the above proposition is also true, because its weak saturation condition implies that Ω is universal, which in turn implies that Ω is atomless.

We now turn to the atomless Loeb probability spaces. It was proved in Hoover and Keisler [1984] that all such spaces are saturated. Here is a simple direct proof.

THEOREM 3B.4. *Every atomless Loeb probability space is saturated.*

PROOF. Let $\Omega = (\Omega, \mathcal{A}, \mu)$ be an internal *finitely additive probability space and let $L(\Omega)$ be the corresponding Loeb space. We assume that $L(\Omega)$ is atomless.

$L(\Omega)$ is universal by Theorem 1E.2. In view of the Saturation Theorem, it suffices to show that $L(\Omega)$ has the back and forth property.

Let $x, \bar{x},$ and y be random variables on $L(\Omega)$ with $x \equiv_0 \bar{x}$. For each n, take a finite subset C_n of M such that $x(\omega)$ is within $1/n$ of C_n with probability at least $1 - 1/n$. Let $x_n(\omega)$ be the point in C_n which is closest to $x(\omega)$. Thus $x_n(\omega) = f_n(x(\omega))$ for a Borel function f_n, and x_n converges to x in probability. Now take $\bar{x}_n(\omega) = f_n(\bar{x}(\omega))$. Since $x \equiv_0 \bar{x}$, we have

$$(x, x_0, x_1, \dots) \equiv_0 (\bar{x}, \bar{x}_0, \bar{x}_1, \dots).$$

Moreover, each x_n and each $\bar{x}_n$ is a simple function, and $\bar{x}_n$ converges to $\bar{x}$ in probability. By Proposition 3B.3, for each n there exists y_n such that

$$(x_0, x_1, \dots, x_n, y_n) \equiv_0 (\bar{x}_0, \bar{x}_1, \dots, \bar{x}_n, \bar{y}).$$

By taking liftings of each of these random variables, using the Saturation Principle, and taking the standard part. we obtain a random variable y on Ω such that

$$(x_0, x_1, \dots, y) \equiv_0 (\bar{x}_0, \bar{x}_1, \dots, \bar{y}).$$

Taking limits in probability, it follows that $(x, y) \equiv_0 (\bar{x}, \bar{y})$, and the back and forth property is proved.   ⊣

Here is a brief account of what can said about saturated and homogeneous probability spaces in the light of results in the classical literature on measure algebras.

DEFINITION 3B.5. *The* **measure algebra** *associated with a probability space* $(\Omega. \mathcal{F}. P)$ *is the structure* $(\mathcal{F}/\mathcal{N}, P/\mathcal{N})$ *where* $\mathcal{F}/\mathcal{N}$ *is the quotient Boolean algebra of* $\mathcal{F}$ *modulo the ideal* $\mathcal{N}$ *of null sets. Two probability spaces are said to be* **measure algebra isomorphic** *if their measure algebras are isomorphic in the natural sense.*

If Ω and Γ are probability measures on disjoint sample sets, $\alpha, \beta > 0$, and $\alpha + \beta = 1$, then the convex combination $\alpha \cdot \Omega + \beta \cdot \Gamma$ is the probability space on $\Omega \cup \Gamma$ formed in the obvious way with Ω and Γ having probabilities α and β. Convex combinations of measure algebras. and countable convex combinations of probability spaces and of measure algebras. are defined in an analogous manner.

Let $[0. 1]^\kappa$ be the probability space formed by taking the product measure of κ copies of the space $[0, 1]$ with the Lebesgue measure. The measure algebras of the spaces $[0, 1]^\kappa$ are of special importance, and are called **homogeneous measure algebras**. The fundamental theorem about measure algebras in Maharam [1942] shows that there are very few measure algebras.

THEOREM 3B.6. (*Maharam's Theorem*) *For every atomless probability space* Ω. *there is a finite or countable set of distinct cardinals* $\{\kappa_i : i \in I\}$ *such that the measure algebra of* Ω *is a convex combination of the homogeneous measure algebras* $[0. 1]^{\kappa_i}$.   ⊣

The set of cardinals κ_i in Maharam's theorem is clearly unique. This set is called the **Maharam spectrum** of Ω. The Maharam spectrum of an ordinary probability space is a set of finite or countable cardinals. We refer the reader to the paper Jin and Keisler [2000] for a discussion of the Maharam spectra of the atomless Loeb probability spaces.

Maharam's Theorem leads to the following characterization of saturated probability spaces.

THEOREM 3B.7. *A probability space is saturated if and only if its Maharam spectrum is a set of uncountable cardinals.*

PROOF. Let Ω be the given probability space. For simplicity we give the proof only for the case that the Maharam spectrum of Ω is a single cardinal κ. The general case is similar.

Suppose κ is countable. Then there is a random variable x on Ω with values in the Polish space $[0, 1]^\kappa$ such that no set of positive measure is independent of x. But in some other probability space there is a pair of random variables (x', y') such that $x' \equiv_0 x$ and y' is the indicator of a set of positive measure which is independent of x'. Then there cannot exist y on Ω such that $(x, y) \equiv_0 (x', y')$, so Ω is not saturated.

Suppose κ is uncountable. Let x be a random variable on Ω, and let (x', y') be a pair of random variables on some other space such that $x \equiv_0 x'$. For some countable subset $J \subseteq \kappa$, x is measurable with respect to the family of sets $A \in \mathcal{F}$ such that $A/\mathcal{N}$ has support J. ⊣

Thus for each uncountable cardinal κ, $[0, 1]^\kappa$ is an example of a saturated probability space which is obtained without nonstandard methods.

COROLLARY 3B.8. *If Ω is a saturated probability space, then any other probability space which is measure algebra is isomorphic to Ω is also saturated.* ⊣

We now consider universal homogeneous probability spaces. The next theorem shows that a universal homogeneous probability space has a homogeneous measure algebra with an uncountable exponent κ.

THEOREM 3B.9. *Suppose Ω is a universal homogeneous probability space. Then the Maharam spectrum of Ω is a single uncountable cardinal.*

PROOF. If the Maharam spectrum contains a finite or countable cardinal, then Ω is not saturated by Theorem 3B.7, so Ω is not homogeneous by the Saturation Theorem.

Suppose the Maharam spectrum contains two distinct cardinals κ, λ. Then Ω has disjoint sets A, B of positive measure such that the restriction of Ω to A and B are measure algebra isomorphic to $[0, 1]^\kappa$ and $[0, 1]^\lambda$ respectively. We may assume without loss of generality that $P[A] = P[B]$. Then $\mathbb{I}_A \equiv_0 \mathbb{I}_B$, but there is no almost sure bijection of Ω which carries $\mathbb{I}_A$ to $\mathbb{I}_B$. This contradicts the hypothesis that Ω is homogeneous as a probability space. Therefore the Maharam spectrum consists of a single uncountable cardinal. ⊣

EXAMPLE 3B.10. *There is a saturated probability space such that neither the space nor its measure algebra is homogeneous.*

PROOF. Let κ and λ be distinct uncountable cardinals. Take a convex combination of the probability spaces $[0, 1]^\kappa$ and $[0, 1]^\lambda$, and apply Theorems 3B.7 and 3B.9.   ⊣

EXAMPLE 3B.11.   *The converse of Theorem 3B.9 fails.*

PROOF. Let κ be an uncountable cardinal, $\lambda > \kappa$, and $\Omega = [0, 1]^\kappa$. Let Γ be the product of Ω with a trivial probability space $(U, \mathcal{G}, Q)$ where $|U| = \lambda$ and $\mathcal{G} = \{U, \emptyset\}$. Then any convex combination of Ω and Γ is saturated and has a homogeneous measure algebra, but is not homogeneous as a probability space because all sets of positive measure have cardinality κ in Ω and λ in Γ.   ⊣

Finally, we give a classical example of a universal homogeneous probability space.

THEOREM 3B.12. *For each uncountable cardinal κ, $[0, 1]^\kappa$ is a universal homogeneous probability space.*

PROOF. By Theorem 3B.7, $[0, 1]^\kappa$ is saturated. Let x and y be M-valued random variables on $[0, 1]^\kappa$ such that $x \equiv_0 y$. Let $U_n, n \in \mathbb{N}$ be a countable open basis for M. Let A be the measure algebra of $[0, 1]^\kappa$, and for each set $J \subseteq \kappa$, let $A(J)$ be the measure algebra of all $a \in A$ with support included in J.

Claim: There is an automorphism f of the measure algebra A which sends $x^{-1}(U_n)$ to $y^{1-}(U_n)$ for each n.

Once this claim is proved, we apply another theorem of Maharam [1958], which says that any automorphism of the measure algebra of $[0, 1]^\kappa$ is realized by an automorphism g of the probability space $[0, 1]^\kappa$. Since f maps $x^{-1}(U_n)$ to $y^{-1}(U_n)$ for each n, we have $x(\omega) = y(g(\omega)) a.s.$, as required.

It remains to prove the Claim. There is a countable $J_0 \subseteq \kappa$ such that $x^{-1}(U_n) \in A(J_0)$ for each n. Since $[0, 1]^\kappa$ is saturated, it follows that there is a measure algebra isomorphism from f_0 from $A(J_0)$ into A such that $f_0(x^{-1}(U_n)) = y^{-1}(U_n))$ for each n. There is a countable J_1 such that $J_0 \subseteq J_1 \subseteq \kappa$ and f_0 maps $A(J_0)$ into $A(J_1)$. Continuing the process back and forth countably many times and taking the union, we obtain a countable set $J = \bigcup_n J_n \subseteq \kappa$ such that $f = \bigcup_n f_n$ is a measure algebra automorphism of $A(J)$. f can then be trivially extended to a measure algebra automorphism of A, and the Claim is proved.   ⊣

3C. Saturated adapted spaces

We now return to the study of saturated adapted spaces.

The following useful theorem is the adapted analog of the Robinson consistency theorem in first order logic. In first order model theory this theorem has an easy proof using recursively saturated models, see Chang and Keisler [1990], Theorem 2.4.8. Here something similar can be done using the universality and homogeneity of the hyperfinite adapted spaces, and the proof is a good exercise. Instead, we will now prove the result directly from the existence of saturated adapted spaces.

THEOREM 3C.1. (*Amalgamation Property, or Robinson Consistency Theorem*). *Let x and y be processes on an adapted space Γ_1 and x' and y' be processes on an adapted space Γ_2 such that $x \equiv x'$. Then there exists an adapted space Ω and processes u, v, w on Ω such that $(x, y) \equiv (u, v)$ and $(x', y') \equiv (u, w)$.*

PROOF. Let Ω be a saturated adapted space. By saturation find processes u and v on Ω such that $(x, y) \equiv (u, v)$. Then $u \equiv x'$, and applying saturation again one can find a process w on Ω such that $(u, w) \equiv (x', y')$.   ⊣

Given the nice behavior of the hyperfinite adapted spaces, it is natural to wonder if there are saturated adapted spaces other than the hyperfinite adapted spaces. The answer is "Yes." We will give two ways of constructing saturated adapted spaces later on this book.

In Hoover and Keisler [1984] (with a correction in Hoover [1990]), methods of Maharam [1950] were used to prove that all atomless Loeb adapted spaces are saturated. In Chapter 7 we will give a completely different proof of this fact, based on Fajardo and Keisler [1996b] and Keisler [1977]. This result will depend on the notion of an atomless adapted space, which is stronger than the notion of an atomless probability space and will also be introduced in Chapter 7.

Hoover [1992] provided a general construction that starts with an arbitrary adapted space and constructs an extension which is saturated. His work is a counterpart to another classical model theoretic construction, that of elementary chains, which we will study in detail in Chapter 6.

At this point, given the results and methods introduced in this chapter, we can ask some natural questions. The first of these was already posed in Fajardo [1987] (Question 2, page 324):

How can we use the properties of hyperfinite adapted spaces within the general theory of stochastic processes?

The papers Hoover and Keisler [1984], Hoover [1984], Hoover [1987], and Barlow and Perkins [1989], Barlow and Perkins [1990] give some applications of saturated adapted spaces in stochastic analysis. The result that hyperfinite adapted spaces are saturated tells us that, at least under the relation $\equiv$, we lose nothing by working with hyperfinite adapted spaces. Going further, the paper Henson and Keisler [1986] suggests that there is much to be gained.

Suppose that we are interested in solving an existence problem in a particular adapted space which is not hyperfinite. Is there a way to use hyperfinite spaces in this case?

One obvious intuitive answer is that if we show that the problem cannot be solved within the hyperfinite space then it cannot be solved in the original space. But what if it can be solved in the hyperfinite space? It may be that we are making essential use of the Saturation Principle for nonstandard universes, and according to Henson and Keisler [1986] this could mean that the problem has no solution in the original space.

Given this situation we would like to identify those problems for which the following informal method works: "Take your problem, translate it to a hyperfinite property, solve it inside the hyperfinite space using its nice combinatorial properties, and then pull your hyperfinite solution back to a solution on your original space." The problem of when this method works is called the **Come-back Problem**.

The above results give a good illustration of the sort of answer that we expect for the Come-back Problem. We hope that more results along these lines will be obtained as more people get to know the underlying ideas of the model theory of stochastic processes. Later on in these notes we will see more results of this type.

We now present one more result which suggests other possible uses of saturated adapted spaces.

PROPOSITION 3C.2. *Let x and y be right continuous processes on adapted spaces Γ and Λ respectively. Suppose that τ and σ are stopping times such that $(x, \tau) \equiv (y, \sigma)$. Then $x^\tau \equiv y^\sigma$.*

PROOF. The proof is by induction on stopping times. That is, we first prove the result for simple stopping times and then, since we know that an arbitrary stopping time is a pointwise decreasing limit of a sequence of simple stopping times, we show that the property is preserved under the operation of pointwise decreasing limits. The first step is suggested in Exercise 1D.8 (b). If the reader has not done that exercise yet, this is a good time. The second step of the induction given here is a "soft" argument due to Carlos A. Sin.

Let Ω be a saturated adapted space. By saturation we can find a right continuous process z and a stopping time θ on Ω such that $(x, \tau) \equiv (z, \theta) \equiv (y, \sigma)$. We prove that $x^\tau \equiv z^\theta \equiv y^\sigma$. We will give one half of the proof chain; the other is identical. Since τ is a stopping time on Γ there is a decreasing sequence (τ_n) of simple stopping times which converges to τ. Since Ω is saturated we can obtain a decreasing sequence (θ_n) of simple stopping times converging to θ and such that

$$(x, \tau, \tau_1, \ldots, \tau_n, \ldots) \equiv (z, \theta, \theta_1, \ldots, \theta_n, \ldots).$$

By right continuity we have that $x^{\tau_n} \to x^\tau$ and $z^{\theta_n} \to z^\theta$. Since for each n, $x^{\tau_n} \equiv z^{\theta_n}$, we get $x^\tau \equiv z^\theta$ as we wanted to prove.   ⊣

The proof we just sketched also exemplifies other uses of the model theoretic concept of saturation. We proved a result that holds for arbitrary adapted spaces by going through a saturated space. By looking at the proof we can ask ourselves additional questions:

(1) *Does the result hold with the relation $\equiv_1$ in place of $\equiv$?*

(2) *Does there exist an adapted space which is saturated with respect to the relation $\equiv_1$?*

If the answer to question (2) were affirmative, then the proof given for Proposition 3C.2 would work again for question (1). Unfortunately, as we shall see later, there in no adapted space which is saturated for the relation $\equiv_1$.

Questions (1) and (2) lead us to another question. Proposition 1D.6 states that the properties of being a local martingale and semimartingale are preserved under the equivalence relation $\equiv_1$, for all adapted spaces. The proof of this result in Hoover [1984] made heavy use of theorems and concepts from the general theory of stochastic processes. Can this be proved by a soft model theoretic argument, at least for the relation $\equiv$?

3D. Adapted density

In this section we prove some technical results that will often let us restrict attention to dense sets of adapted functions and conditional processes, or to cadlag (right continuous left limit) stochastic processes. These results are proved with nontrivial measure theoretic arguments (see Hoover and Keisler [1984], Section 2, and Keisler [1984], Lemma 3.1).

DEFINITION 3D.1. (a) *Let F be a set of adapted functions. F is said to be $\equiv$-dense if for any stochastic process y and sequence (x_n) of stochastic processes (possibly on different adapted spaces), if $E[f(x_n)] \to E[f(y)]$ for all $f \in F$ then $E[f(x_n)] \to E[f(y)]$ for every adapted function f.*

(b) *Let H be a function which assigns to each Polish space M a set $H(M)$ of bounded continuous functions from M into $\mathbb{R}$. The set $F(H, M)$ of adapted functions built from $H(M)$ is the set of adapted functions built using only elements from the set $H(M)$ at the Basis Steps and using only polynomials with rational coefficients at the Composition Steps.*

Observe that if F is $\equiv$-dense, then whenever $E[f(x)] = E[f(y)]$ for all $f \in F$ we have $x \equiv y$. However, the property that F is $\equiv$-dense is actually stronger than the property that $x \equiv y$ whenever $E[f(x)] = E[f(y)]$ for all $f \in F$. See the discussion of convergence determining sets in Ethier and Kurtz [1986], for example.

PROPOSITION 3D.2. *One can choose for each Polish space M a countable set $H(M)$ of bounded continuous functions from M into $\mathbb{R}$ such that the set of adapted functions built from $H(M)$ is $\equiv$-dense.*

PROOF. A set $H(M)$ of bounded functions $f : M \to \mathbb{R}$ is said to be **bounded pointwise dense** if every bounded Borel function $g : M \to \mathbb{R}$ belongs to the closure of $H(M)$ under pointwise convergence of uniformly bounded sequences of functions. It is shown in Ethier and Kurtz [1986], Proposition 4.2, that for each Polish space M there is a countable bounded pointwise dense set $H(M)$ of continuous functions. It follows from Hoover and Keisler [1984], Theorem 2.26 that the set of adapted functions built from H is $\equiv$-dense.   ⊣

In Proposition 3D.2 we can replace adapted functions by conditional processes, and of course it remains true (with the appropriate definition of $\equiv$-dense). The interesting thing is that the set of conditional processes built from $H(M)$ is itself countable! The reason should be clear: in the conditional processes definition of adapted distributions, time is treated as a free variable. In the adapted function approach we might have a different adapted function for each tuple of times, since times are constants.

In the case that the time set $\mathbb{L}$ is countable, the analog of an $\equiv$-dense set of adapted functions is an $\equiv_{\mathbb{L}}$-dense set of $\mathbb{L}$-adapted functions, and Proposition 3D.2 takes a simpler form.

COROLLARY 3D.3. *Let $\mathbb{L}$ be a countable time set. For each Polish space M there is a countable $\equiv_{\mathbb{L}}$-dense set of $\mathbb{L}$-adapted functions.*

PROOF. If $\mathbb{L}$ is countable, then the set of $\mathbb{L}$-adapted functions built from the countable set of continuous functions from Proposition 3D.2 is itself countable.   ⊣

PROPOSITION 3D.4. *Suppose x is right continuous in probability. Then for each conditional process f, $f(x)$ is right continuous in probability.*   ⊣

This result is proved in Hoover and Keisler [1984], as Theorem 2.10. The proof is by a routine induction on conditional processes. It follows from the last two propositions that only countably many adapted functions are needed to establish the adapted equivalence of a pair of cadlag processes.

COROLLARY 3D.5. *Let $\mathbb{L}$ is a dense subset of $[0, 1]$ containing 1.*

(a) If x and y are right continuous in probability and $x \equiv_{\mathbb{L}} y$, then $x \equiv y$.

(b) There is a countable set F of $\mathbb{L}$-adapted functions such that for any two stochastic process x, y which are right continuous in probability, $x = y$ if and only if $E[f(x)] = E[f(y)]$ for all $f \in F$.

PROOF. (a) This follows from the fact that any two right continuous functions on $[0, 1]$ which agree on $\mathbb{L}$ are equal.

(b) By Proposition 3D.2, there is a countable set S of conditional processes which is $\equiv$-dense. Let F be the set of all adapted functions obtained from members of S by replacing the time variables by numbers in $\mathbb{L}$. Then F is countable. By Proposition 3D.4, $E[g(x)]$ and $E[g(y)]$ are right continuous functions of time for each $g \in S$. Thus if $E[f(x)] = E[f(y)]$ for all $f \in F$ then $E[g(x)] = E[g(y)]$ for all $g \in S$ and all times.   ⊣

Note that Proposition 3D.4 and Corollary 3D.2 hold in particular when x, y are random variables.

As an application of the results of this section, we will obtain natural characterizations of universal, homogeneous, and saturated adapted spaces which are stated in terms of the classical notion of a Markov process and do not mention adapted functions. The following definition is from Hoover and Keisler [1984], Definition 2.16, but simplified by restricting our attention to random variables, and using adapted functions rather than conditional processes.

DEFNITION 3D.6. *Let $\mathbb{Q}$ be the set of rationals in $[0, 1]$. (Any other countable dense subset of $[0, 1]$ containing 1 will do as well). For each Polish space M let $H(M)$ be as in Proposition 3D.2, and let $F(H, M)$ be the countable set of $\mathbb{Q}$-adapted functions with values in M built from functions in $H(M)$. For each M-valued random variable x, define mx as the stochastic process with values in $\mathbb{R}^{\mathbb{N}}$*

defined by

$$mx(t) = \langle E[fx|\mathcal{F}_t] : f \in F(H, M) \rangle.$$

The process mx is related to the prediction process in Knight [1975]. The main facts about mx are stated in the next proposition, from Hoover and Keisler [1984], Propositions 2.17, 2.18, and Lemma 5.6.

PROPOSITION 3D.7. *Let x and y be random variables.*

(a) mx is an $\mathcal{F}_t$-Markov process with a right continuous version.

(b) $x \equiv y$ if and only if $mx \equiv_0 my$, and also if and only if $(x, mx) \equiv_0 (y, my)$.

(c) For each Polish space M there is a Borel function $\theta : \mathbb{R}^{\mathbb{N}} \to M$ such that for each M-valued random variable x, $x = \theta(mx(1))$.

PROOF. (a) mx has a right continuous version because each coordinate is an $\mathcal{F}_t$-martingale. It is Markov because for all times $s < t$ with t rational, for all k-tuples $\vec{f}(x)$ of coordinates of mx, and for all k-ary rational polynomials $h : \mathbb{R}^k \to \mathbb{R}$, the process $g_s = E[h(\vec{f}(x)(t))|\mathcal{F}_s]$ is another coordinate of mx.

(b) If $x \equiv y$, then $mx \equiv_0 my$ because each coordinate of mx is $E[f(x)|\mathcal{F}_t]$ for some adapted function f, and the corresponding coordinate of my is $E[f(y)|\mathcal{G}_t]$ with the same adapted function f. If $mx \equiv_0 my$, then $(x, mx) \equiv_0 (y, my)$ because the laws of (x, mx) and (y, my) are determined in the same way by adapted functions of x and y.

If $(x, mx) \equiv_0 (y, my)$, then $mx(1) \equiv_0 my(1)$. The coordinates of $mx(1)$ and $my(1)$ form $\equiv_{\mathbb{Q}}$-dense sets of $\mathbb{Q}$-adapted functions. Thus $x \equiv_{\mathbb{Q}} y$, and $x \equiv y$ by Corollary 3D.5.

(c) The required function θ is the function whose graph is the intersection of all open rectangles $A \times B$ such that for some rational open interval I and for some n, the nth coordinate of $mx(1)$ is a basic adapted function $\phi(x)$ where $\phi \in H(M)$, and

$$A = \{\alpha \in \mathbb{R}^{\mathbb{N}} : \alpha_n \in I\}, \; B = \{p \in M : \phi(p) \in I\}. \qquad \dashv$$

We can now give the promised characterizations, using only the classical notions of a Markov process and finite dimensional distribution.

THEOREM 3D.8. *Let Ω be an adapted space.*

(a) Ω is universal if and only if for every right continuous Markov process x on some adapted space, there is a right continuous Markov process y on Ω with $x \equiv_0 y$.

(b) Ω is homogeneous if and only if for all right continuous Markov processes x, y on Ω such that $x \equiv_0 y$, there is an adapted automorphism of Ω with $y(\omega, t) = x(h(\omega), t)$ a.s.

(c) Ω is saturated if and only if for all right continuous Markov processes x on Ω and (x', y') on some adapted space, there is a right continuous process y on Ω such that (x, y) is Markov and $(x, y) \equiv_0 (x', y')$.

PROOF. (a) Assume Ω is universal and let x be a right continuous Markov process. There exists a process y on Ω with $x \equiv y$. It follows that y has a right continuous version. By Proposition 1D.3, y is a Markov process on Ω.

For the converse, by Lemma 1E.6 it suffices to prove that Ω is universal for random variables. Let x be a random variable on some adapted space. By Proposition 3D.7 (a), mx is a Markov process with a right continuous version. By Proposition 3D.7 (c), $x = \theta(mx(1))$. By hypothesis there is a right continuous Markov process z on Ω with $z \equiv_0 mx$. By Proposition 1D.3, $z \equiv mx$. Let $y = \theta(z(1))$. By Proposition 1D.2 (c) we have $y \equiv x$. This proves (a).

The proofs of (b) and (c) are similar.   ⊣

We also get a characterization of adapted equivalence of random variables which is stated strictly in terms of Markov processes. Another characterization, in terms of adapted processes, is given by Hoover [1987]

THEOREM 3D.9. *Let x, y be random variables on adapted spaces Ω, Γ with values in a Polish space M. The following are equivalent.*

(i) x is adapted equivalent to y, $x \equiv y$.

(ii) There exist $\mathbb{R}^N$-valued right continuous Markov processes u on Ω and v on Γ and a Borel function $\theta : \mathbb{R}^N \to M$ such that

$$u \equiv_0 v, \quad \theta(u(1)) = x, \quad \theta(v(1)) = y.$$

PROOF. Assume (i), and let $u = mx$ and $v = my$. Then by Proposition 3D.7 (a), mx and my are right continuous Markov processes with values in $\mathbb{R}^N$ such that $mx \equiv_0 my$. By Proposition 3D.7 (c). $x = \theta(mx(1))$ and $y = \theta(my(1))$ and θ is Borel. This proves (ii).

Assume (ii). By Proposition 1D.3 we have $u \equiv v$. Then $x \equiv y$ by Proposition 1D.2 (c). and (i) is proved.   ⊣

The above theorem also gives a characterization of $x \equiv y$ for stochastic processes x, y. because $x \equiv y$ if and only if $x(\vec{t}) \equiv y(\vec{t})$ for all finite tuples $\vec{t}$ in $[0, 1]$.

3E. Abstract saturation

Given the results that we have obtained so far. a natural question comes up: *Do we really need the adapted equivalence relation $\equiv$?* Perhaps we can obtain similar or at least "close enough" results with the synonymity relation $\equiv_1$ instead of $\equiv$. As pointed out in Aldous [1981], the relation $\equiv_1$ is very natural to work with. It is simpler than $\equiv$, and still preserves many interesting notions in the theory of general stochastic processes.

In this section we are going to show that in spite of the apparent good behavior of the relation $\equiv_1$, it is not good enough for our purposes. In order to back up this statement we present results that come from Section 5 of the paper Hoover and Keisler [1984], and show that $\equiv$ is the weakest relation that has a certain list of properties.

The results of this section may be considered to belong to the abstract model theory for models with stochastic processes. in the spirit of the theorems of Per

Lindström (see Chang and Keisler [1990]) which characterize first order logic (as suggested in Keisler [1985]).

The results below are in fact slightly weaker than Theorems 5.11 and 5.12 in Hoover and Keisler [1984]. We have chosen to treat them in a way that fits in naturally with our previous results. We first introduce a general notion of saturation relative to an equivalence relation $\approx$.

DEFINITION 3E.1. *Let $\approx$ be an equivalence relation on the class of all stochastic processes. An adapted space Ω will be called* **saturated for** $\approx$ *if for each pair of stochastic processes x, y on some other adapted space Γ, for each stochastic process x' on Ω such that $x' \approx x$ there exists a stochastic process y' on Ω with $(x', y') \approx (x, y)$.*

Note that an adapted space Ω is saturated in our previous sense if and only if it is saturated for $\equiv$. Here is our first theorem on saturation for $\approx$.

THEOREM 3E.2. *Let Ω be an adapted space and $\approx$ be an equivalence relation on the class of all stochastic processes on Ω. Assume that Ω and $\approx$ satisfy the following properties*:

(i) $x \approx y$ *implies* $x \equiv_0 y$.

(ii) *If* $(x_1, x_2) \approx (y_1, y_2)$ *then* $x_2 \approx y_2$.

(iii) *If $x \approx y$ and x is right continuous and adapted, then y is also right continuous and adapted*.

(iv) *Ω is saturated for $\approx$.*

If x and y are stochastic processes on Ω and $x \approx y$, then $x \equiv y$.

The main ingredient in the proof of the theorem is the Markov process mx associated with a random variable x which was defined in the preceding section. We need one more lemma about mx.

LEMMA 3E.3. *Let y be an M-valued random variable defined on an adapted space Γ. Let z be a right continuous stochastic process on an adapted space Ω such that $z \equiv_0 my$ and each coordinate of z is a martingale. Then z is a version of mx where $x = \theta(z(1))$.*

PROOF. The proof that mx is a Markov process in Proposition 3D.7 (a) goes through under the present hypotheses to show that z is a Markov process. Therefore by Proposition 1D.3, $z \equiv my$. By Proposition 3D.7 (c), $y = \theta(my(1))$. Then $(x, z) \equiv (y, my)$ by Proposition 1D.2 (c). It follows that $z(t) = mx(t)$ a.s. for each t, that is, z is a version of mx. ⊣

PROOF. of Theorem 3E.2. Suppose first that x, y are random variables on Ω with values in some Polish space M and that $x \approx y$. We claim that $x \equiv y$ in this case. By hypothesis (iv), there is a stochastic process z on Ω such that $(x, z) \approx (y, my)$. By hypotheses (ii), $z \approx my$. By hypothesis (iii), z is adapted, and by hypothesis (i) we have $(x, z) \equiv_0 (y, my)$. By hypothesis (ii), $z_n \approx (my)_n$ for each coordinate n.

We claim that each coordinate z_n is a martingale. Suppose not. Then there are times $r < s$ in $\mathbb{Q}$ and a set $A \in \mathcal{F}_r$ such that $E[z_n(r) \cdot \mathbb{I}_A] \neq E[z_n(s)\mathbb{I}_A]$. Now consider the stochastic process u_t on Ω defined by

$$u_t(\omega) = 0 \text{ if } t < r \text{ and } u_t(\omega) = \mathbb{I}_A(\omega) \text{ if } t \geq r.$$

The process u is adapted. Since Ω is saturated for $\approx$ we can find a stochastic process v on Ω such that $(z, u) \approx (my, v)$. and by hypothesis (ii) again, $u \approx v$. Since the relation $\approx$ preserves adaptedness by hypothesis (iii), the stochastic process v is adapted. Applying saturation for $\approx$ and hypothesis (ii) once more, we get a subset $B \subseteq \Omega$ such that $(u, \mathbb{I}_A) \approx (v, \mathbb{I}_B)$. By hypothesis (i), we have

$$(u, \mathbb{I}_A) \equiv_0 (v, \mathbb{I}_B).$$

Therefore

$$v_t(\omega) = 0 \text{ if } t < r \text{ and } v_t(\omega) = \mathbb{I}_B(\omega) \text{ if } t \geq r \ a.s.$$

(Notice that we have to add a.s. here). Since v is adapted and $A \in \mathcal{F}_r$, it follows that $B \in \mathcal{F}_r$. Observe that

$$E[(my)_n(r) \cdot \mathbb{I}_B] = E[z_n(r) \cdot \mathbb{I}_A] \neq E[z_n(s) \cdot \mathbb{I}_A] = E[(my)_n(s) \cdot \mathbb{I}_B].$$

The two ends of the above line contradict the fact that $(my)_n$ is a martingale. This proves the claim.

It now follows from Lemma 3E.3 (b) that $z = mx$, so that

$$(y, my) \equiv_0 (x, mx).$$

Hence by Proposition 3D.7, $x \equiv y$. We have completed the proof in the case that x, y are random variables.

Now let x and y be stochastic processes on Ω such that $x \approx y$. By Lemma 1E.3 there is a sequence $(t_n)_{n \in \mathbb{N}}$ of elements in $[0, 1]$ and two Borel functions $\psi : [0, 1] \to 2^{\mathbb{N}}, \phi : 2^{\mathbb{N}} \to M$ such that for all $t \in [0, 1]$,

$$x(\cdot, t) = \phi(\psi(t, \langle x(\cdot, t_n) : n \in \mathbb{N} \rangle)) \ a.s.$$

By saturation for $\approx$ there is a random variable v on Ω such that

$$(x, \langle x(\cdot, t_n) : n \in \mathbb{N} \rangle) \approx (y, v).$$

By hypothesis (i),

$$(x, \langle x(\cdot, t_n) : n \in \mathbb{N} \rangle) \equiv_0 (y, v).$$

Therefore for all t, $y(\cdot, t) = \phi(\psi(t, v)) \ a.s.$ By hypothesis (ii) we have

$$\langle x(\cdot, t_n) : n \in \mathbb{N} \rangle \approx v.$$

By the random variable case which we have already proved,

$$\langle x(\cdot, t_n) : n \in \mathbb{N} \rangle \equiv v.$$

By Proposition 1D.2 (c),

$$x(t) \equiv \phi(\psi(t, v)) = y(t) \ a.s.$$

for all t. and hence $x \equiv y$.   ⊣

DEFINITION 3E.4. *An equivalence relation $\approx$ on the class of all stochastic processes has the* **amalgamation property** *if for all stochastic processes x and y on an adapted space Γ_1, and x' and y' on an adapted space Γ_2, if $x \approx x'$ then there exists an adapted space Ω and processes u, v, w on Ω such that $(x, y) \approx (u, v)$ and $(x', y') \approx (u, w)$.*

Thus an equivalence relation $\approx$ has the amalgamation property if and only if Corollary 3C.1 holds with $\approx$ in place of $\equiv$. The proof of that corollary actually shows more.

COROLLARY 3E.5. *If $\approx$ is an equivalence relation on the class of all stochastic processes and there exists an adapted space which is saturated for $\approx$, then $\approx$ has the amalgamation property.*   ⊣

The following theorem is similar in spirit to Theorem 3E.2. We omit the proof because it is essentially subsumed in the proof of Theorem 3E.2.

THEOREM 3E.6. *Let $\approx$ be an equivalence relation on the class of all stochastic processes, which satisfies the following properties:*

(i) $x \approx y$ implies $x \equiv_0 y$.

(ii) If $(x_1, x_2) \approx (y_1, y_2)$ then $x_2 \approx y_2$.

(iii) If $x \approx y$ and x is a martingale then y is also a martingale.

(iv) $\approx$ has the amalgamation property.

Then $x \approx y$ implies $x \equiv y$.   ⊣

COROLLARY 3E.7. *The relation $\equiv_1$ does not have the amalgamation property. Moreover, no adapted space is saturated for $\equiv_1$. The same is true of all the $\equiv_n$.*

PROOF. Clearly $\equiv_1$ satisfies hypotheses (i)–(iii) of Theorem 3E.6. Proposition 1D.5 tells us that there are processes x and y with $x \equiv_1 y$ but not $x \equiv y$, so the conclusion of Theorem 3E.6 fails for $\equiv_1$. Therefore $\equiv_1$ cannot satisfy hypothesis (iv) of Theorem 3E.6, that is, $\equiv_1$ cannot have the amalgamation property. Finally, by Corollary 3E.5, no adapted space is saturated for $\equiv_1$.   ⊣

Given the way saturated adapted spaces are used in stochastic analysis, the above corollary is definitely a good argument against the relation $\equiv_1$. In the following chapters we are going to present other arguments. But before we do that, let us call attention to a question posed in Keisler [1985] which comes up naturally at this point and leads us to the subject of the next chapter.

QUESTION 3E.8. *Is there an equivalence relation on stochastic processes which satisfies the hypotheses (i)-(iv) of Theorem 3E.6, and is stronger than $\equiv$ but weaker than the adapted isomorphism relation $\simeq$?*

The answer will be given in the next chapter, in Section 4C.

CHAPTER 4

COMPARING STOCHASTIC PROCESSES

Are there other ways of comparing stochastic processes on adapted spaces besides those which we have considered so far? The question posed at the end of the last chapter suggests that there are. In this chapter we will introduce equivalence relations between stochastic processes which extend the relation $\equiv$. By comparing these new equivalence relations with $\equiv$, we will get a better understanding of the strengths of the relation $\equiv$, and at the same time discover some of its weaknesses.

Section 4A will be devoted to results of from Hoover [1987] which show how, motivated by a question in stochastic analysis, one can compare the strength of the relation $\equiv$ to the relations $\equiv_n$. Then, in the rest of the chapter, we will combine ideas from Hoover [1987], Fajardo [1990a] and Fajardo and Peña [1997] to obtain some interesting new ways of comparing stochastic processes.

4A. Finite embeddability

We begin this section by introducing Hoover's notion of finite embeddability between stochastic processes.

DEFNITION 4A.1. *Let x and y be stochastic processes on adapted spaces Ω and Γ respectively. (Ω, x) is said to be **finitely embeddable** in (Γ, y), in symbols $(\Omega, x) \to_0 (\Gamma, y)$ or $x \to_0 y$, if the following property holds:*
For all $k \in \mathbb{N}$, all $t_1 < \cdots < t_k$ in $[0, 1]$, and all random variables $x_1, \ldots, x_k$ on Ω which are measurable with respect to the σ-algebras $\mathcal{F}_{t_1}, \ldots, \mathcal{F}_{t_k}$, there are random variables $y_1, \ldots, y_k$ on Γ measurable with respect to the σ-algebras $\mathcal{G}_{t_1}, \ldots, \mathcal{G}_{t_k}$ such that

$$(x, x_1, \ldots, x_k) \equiv_0 (y, y_1, \ldots, y_k).$$

Let us write $x \leftrightarrow_0 y$ if both $x \to_0 y$ and $y \to_0 x$. It is clear that the finite embeddability relation $\to_0$ is transitive, and that $\leftrightarrow_0$ is an equivalence relation.

Here is the main theorem concerning this new equivalence relation.

THEOREM 4A.2. *If x and y are stochastic processes and $x \leftrightarrow_0 y$ then $x \equiv y$.*

In the course of the proof of the theorem we will need the following technical fact that is best proved separately.

LEMMA 4A.3. *Let x and y be square integrable random variables on probability spaces $(\Lambda, \mathcal{F}, P)$ and $(\Gamma, \mathcal{G}, Q)$ respectively. Suppose $\mathcal{F}_0$ and $\mathcal{G}_0$ are σ-algebras with $\mathcal{F}_0 \subseteq \mathcal{F}$ and $\mathcal{G}_0 \subseteq \mathcal{G}$, and z and w are random variables measurable with respect to $\mathcal{F}_0$ and $\mathcal{G}_0$ respectively. If*

$$(x, E[x|\mathcal{F}_0]) \equiv_0 (y, w)$$

and

$$(x, z) \equiv_0 (y, E[y|\mathcal{G}_0])$$

then

$$z = E[x|\mathcal{F}_0] \ a.s. \ and \ w = E[y|\mathcal{G}_0] \ a.s.$$

PROOF. Let us denote the random variable $E[x|\mathcal{F}_0]$ by x' and $E[y|\mathcal{G}_0]$ by y'. Observe that by Jensen's inequality for conditional expectations, both x' and y' are square integrable. Since this property is preserved under $\equiv_0$, from the hypothesis we have that z and w are also square integrable.

Using basic properties of conditional expectations it is easy to see that the following properties are true: $E[x] = E[x']$, $E[zx'] = E[zx]$, and $E[(x')^2] = E[x'x]$. Similarly for y, y' and w. Letting σ^2 denote variance, we have

(5)
$$\sigma^2(z - x) = \sigma^2(z - x') + \sigma^2(x' - x), \sigma^2(w - y) = \sigma^2(w - y') + \sigma^2(y' - y).$$

By hypothesis,

(6) $$\sigma^2(z - x) = \sigma^2(y' - y) \text{ and } \sigma^2(w - y) = \sigma^2(x' - x).$$

Then if we add both sides of (5) and substitute the values of (6) we obtain

$$\sigma^2(z - x') + \sigma^2(w - y') = 0.$$

Consequently $z = x'$ *a.s.* and $w = y'$ *a.s.*. as we wanted to show.   ⊣

A curiosity: In the above proof, the fact that the random variables were square integrable was used very heavily, since we needed their variances. Is the result true without this hypothesis?

PROOF. of Theorem 4A.2: Assume that $x \leftrightarrow_0 y$. Given an adapted function f, let $Sub(f)$ be the finite set of adapted functions used in the definition of f.

For each $g \in Sub(f)$, $g(x)$ and $g(y)$ are random variables which are $\mathcal{F}_1$ and $\mathcal{G}_1$ measurable. If g is of the form $E[h|t]$ then $g(x)$ and $g(y)$ are $\mathcal{F}_t$ and $\mathcal{G}_t$ measurable. By hypothesis, there exist random variables $y^g, g \in sub(f)$ such that

$$(x, g(x))_{g \in Sub(f)} \equiv_0 (y, y^g)_{g \in Sub(f)}$$

and y^g is $\mathcal{G}_t$ measurable whenever $g(x)$ is $\mathcal{F}_t$ measurable. Similarly, there exist $x^g, g \in Sub(f)$ such that

$$(x, x^g)_{g \in Sub(f)} \equiv_0 (y, g(y))_{g \in Sub(f)}$$

and x^g is $\mathcal{F}_t$ measurable whenever $g(y)$ is $\mathcal{G}_t$-measurable.

We now prove a property from which the theorem follows right away: For each adapted function $g \in Sub(f)$, $g(x) = x^g$ *a.s.* and $g(y) = y^g$ *a.s.* This is done by induction on the complexity of g. The basis and composition steps are easy and the conditional expectation step is exactly what the previous lemma is all about. Simply observe that by definition, the interpretation of an adapted function is always uniformly bounded and hence square integrable.   ⊣

In his paper Hoover also introduced another relation, which is a priori the same or weaker than $x \rightarrow_0 y$.

DEFINITION 4A.4. *Let us write* $x \Rightarrow_0 y$ *if the condition in Definition 4A.1 holds for simple random variables* x_i *and* y_i. *That is, for all k, $t_1 < \cdots < t_k$, and simple random variables* $x_1, \ldots, x_k$ *with* x_i *$\mathcal{F}_{t_i}$-measurable, there exist simple random variables* $y_1, \ldots, y_k$ *measurable with respect to the σ-algebras* $\mathcal{G}_{t_1}, \ldots, \mathcal{G}_{t_k}$ *such that*

$$(x, x_1, \ldots, x_k) \equiv_0 (y, y_1, \ldots, y_k).$$

Equivalently, we could require the x_i and y_i to be characteristic functions of measurable sets A_i, B_i taken from the filtrations, such that

$$(x, \mathbb{I}_{A_1}, \ldots, \mathbb{I}_{A_k}) \equiv_0 (y, \mathbb{I}_{B_1}, \ldots, \mathbb{I}_{B_k}).$$

It is clear that $x \rightarrow_0 y$ implies $x \Rightarrow_0 y$. Following the pattern for $\leftrightarrow_0$, we write $x \Leftrightarrow_0 y$ if both $x \Rightarrow_0 y$ and $y \Rightarrow_0 x$. Hoover also proved the statement of Theorem 4A.2 with $\Leftrightarrow_0$ in place of $\leftrightarrow_0$. For future reference we state the result here. The proof is very similar to that of Theorem 4A.2; for details see Hoover [1987].

THEOREM 4A.5. *If x and y are stochastic processes and $x \Leftrightarrow_0 y$ then $x \equiv y$.*   ⊣

An easy corollary of the above theorem is that if x and y are processes on saturated adapted spaces we have:

$$x \equiv y \text{ if and only if } x \Leftrightarrow_0 y.$$

Is this true in general? No. The new relation $x \Leftrightarrow_0 y$ conveys extra information that is independent of the original processes x, y and has to do with the richness of the filtrations. This observation suggests a very simple counterexample.

EXAMPLE 4A.6. *Let Γ be a finite adapted space and let y be a stochastic process on Γ. If Ω is a hyperfinite adapted space, by universality there exists a process x on Ω such that $x \equiv y$. Obviously we cannot have $x \Leftrightarrow_0 y$.*

Hoover [1987] considered the following question.

QUESTION 4A.7. *Does $x \Leftrightarrow_0 y$ imply $x \leftrightarrow_0 y$?*

Hoover [1987] conjectured a positive answer, but as we will see later on in this chapter, Peña [1993] proved the answer is negative.

In order to illustrate how relations of this type come up in the model theory of stochastic processes, we will describe the problem that prompted Hoover to consider the relations $\leftrightarrow_0$ and $\Leftrightarrow_0$, and give a brief sketch of his solution. It is important to realize that it was a problem in stochastic analysis that sparked Hoover's work and led him to use tools from the model theory of stochastic processes. Some of the statements depend on advanced notions from stochastic analysis which we will make no attempt to explain here. We will state some of

Hoover's results without proof, since our purpose at this point is to illustrate Hoover's method without working out the details. We first need a definition.

DEFINITION 4A.8. *Let u and v be semimartingales. The following property is denoted by* $u \rightarrow_{sde} v$. *Whenever* f *is a measurable function from* $[0, 1] \times \mathbb{R}$ *into* $\mathbb{R}$ *and there exist processes* $x^1, \ldots, x^k$ *satisfying*

$$(7) \qquad\qquad x_t^i = x_0^i + \int_0^t f(s, x_{s-}^i) \, du_s$$

then there are processes $y^1, \ldots, y^k$ *satisfying* (7) *with the* y^i's *in place of the* x^i's, *such that*

$$(u, x^1, \ldots, x^k) \equiv_0 (v, y^1, \ldots, y^k).$$

As usual, $u \leftrightarrow_{sde} v$ means $u \rightarrow_{sde} v$ and $v \rightarrow_{sde} u$.

The definition of $u \rightarrow_{sde} v$ gives us a good idea of why Hoover considered the relations $\Leftrightarrow_0$ and $\leftrightarrow_0$. The relation $\leftrightarrow_{sde}$ can be used to give concise statements of some previously known results. For example, Theorem 7.8 in Hoover and Keisler [1984] can be stated as follows.

THEOREM 4A.9. *Let u and v be semimartingales on a saturated adapted space* Ω. *If* $u \equiv v$ *then* $u \leftrightarrow_{sde} v$. $\dashv$

Hoover said in the introduction of his paper Hoover [1987] that his work began when he tried to improve the above result by replacing $\equiv$ by $\equiv_1$. But instead, he was able to prove the following result.

THEOREM 4A.10. *Let u and v be continuous local martingales with strictly increasing quadratic variation (and hence semimartingales).*
  (a) *If* $u \rightarrow_{sde} v$ *then* $u \Rightarrow_0 v$.
  (b) *If* $u \leftrightarrow_{sde} v$ *then* $u \equiv v$.
  (c) $\equiv$ *cannot be replaced by* $\equiv_1$ *in Theorem 4A.9.*

PROOF. We just give a proof of (c). (Hoover did not write the proof explicitly, and it is not clear that the following argument is the one that he has in mind).

Let u and v be continuous local martingales with strictly increasing quadratic variations (on some adapted spaces) such that $u \equiv_1 v$ but not $u \equiv v$. (You can ask yourself: Are there u and v which satisfy this condition?). Find u' and v' which are continuous local martingales with strictly increasing quadratic variations on a saturated adapted space Ω such that $u \equiv u'$ and $v \equiv v'$. Then $u' \equiv_1 v'$ but not $u' \equiv v'$. Therefore by (b), we have $u' \equiv_1 v'$ but not $u' \leftrightarrow_{sde} v'$. $\dashv$

4B. Stronger equivalence relations

In the paper "Games between adapted spaces", Fajardo [1990a] answered the question posed at the end of the previous chapter: Are there relations that properly extend $\equiv$ and are strictly weaker than $\simeq$? In order to do this a new way of comparing stochastic processes was introduced. So before we give the solution given in Fajardo [1990a], we are going to introduce these new relations and explore some of their properties.

Even though the results from Fajardo [1990a] were not motivated by Hoover [1987], we can present them in a closely related way. In his master's thesis, Peña [1993] answered several questions related to problems posed by Hoover [1987] and Fajardo [1990a]. These results and more appear in the paper Fajardo and Peña [1997].

A well-known method for comparing structures in model theory is the back-and-forth method (see Hodges [1985]), which is sometimes phrased in terms of the Ehrenfeucht-Fraisse game. We are going to do something similar here. In fact, many of the ideas contained in Fajardo [1990a] and Fajardo [1987] were motivated by trying to give a back-and-forth characterization of $\equiv$.

Intuitively, $(\Omega, x) \Leftrightarrow_0 (\Gamma, y)$ means that the spaces Ω and Γ have, at all times, the same distributions of finite families of simple random variables relative to x and y. Similarly, $(\Omega, x) \leftrightarrow_0 (\Gamma, y)$ means that the spaces Ω and Γ have, at all times, the same finite families of random variables relative to x and y.

We will define stronger relations by slightly modifying the rules for $\Leftrightarrow_0$ and $\leftrightarrow_0$. There will always be a simple random variable version (i.e., choosing sets) which we denote with a $\Leftrightarrow$ symbol, and a general random variable version with a $\leftrightarrow$ symbol. Again, in each case we must consider the problem of whether or not the relations are different. Here is a list of definitions with comments and questions. These relations were studied in Fajardo [1990a], Fajardo and Peña [1997] and Peña [1993].

In Definition 4A.1 above, we chose the times at which the adapted sets are picked and then required $\equiv_0$-equivalence. We can ask what would happen if instead of $\equiv_0$-equivalence we required $\equiv_1$, $\equiv_n$, or $\equiv$-equivalence? Let us denote the simple random variable forms of these equivalence relations by $\Leftrightarrow_1$, $\Leftrightarrow_n$ and $\Leftrightarrow_\infty$. The corresponding arbitrary random variable versions of these new relations are denoted by $\leftrightarrow_1$, $\leftrightarrow_n$ and $\leftrightarrow_\infty$. (The n indicates what type of $\equiv_n$ equivalence is used in the definition, and ∞ indicates that the full adapted equivalence relation $\equiv$ is used). Clearly, the equivalence relations get stronger as n increases.

The one-directional relations $\rightarrow_n$ and $\Rightarrow_n$ are also defined in the obvious way. Moreover, all of these relations can be considered, with no change in the definitions, between tuples of stochastic processes x and y, or even between stochastic processes with values in a Polish space M.

When $n > 0$, we do not have to say at which times the adapted sets are chosen, since two sets that are at least synonymous (i.e., at least $\equiv_1$-equivalent) are adapted at the same times. Thus in the relations $\Leftrightarrow_n$ or $\leftrightarrow_n$ when $n > 0$, we say that for each finite tuple of sets or random variables on Ω there is a tuple on Γ of the same length such that . . . .

A trivial observation lets us see what can be expected. In our original relation $\Leftrightarrow_0$ one may have

$$(x, \mathbb{I}_{A_1}, \dots, \mathbb{I}_{A_k}) \equiv_0 (y, \mathbb{I}_{B_1}, \dots, \mathbb{I}_{B_k})$$

with sets $A_1 \in \mathcal{F}_t$ and $B_1 \in \mathcal{G}_t$, such that B_1 also belongs to $\mathcal{G}_s$ for an s with $t < s$ but A_1 does not belong to $\mathcal{F}_s$. There is nothing in the definition of $\Leftrightarrow_0$ that could prevent this situation. With the $\equiv_n$'s or $\equiv$ this could never happen. A new set of problems arise.

PROBLEM 4B.1. (a) Do we have, in analogy with the $\equiv_n$'s, that the relations $\Leftrightarrow_0, \Leftrightarrow_1, \ldots, \Leftrightarrow_\infty$ are all different? (See Proposition 1D.5). Similarly for $\leftrightarrow_0, \leftrightarrow_1$, ..., $\leftrightarrow_\infty$.

(b) Do any of these relations have the amalgamation property?

A "Yes" answer to both (a) and (b) would solve Problem 3E.8.

A natural way to modify the $\to_n$ or $\leftrightarrow_n$ relation is to choose infinite sequences of sets or random variables, or even to choose random variables with values in an arbitrary Polish space. The next theorem, which is a reformulation of a result of [P], shows that when $n > 0$ these modifications will make no difference. In fact, it is enough to choose a single random variable.

THEOREM 4B.2. Let $n > 0$. Let x be a stochastic process on Ω and y a stochastic process on Γ. The following are equivalent, with $\equiv$ in place of $\equiv_n$ in the case $n = \infty$.

(i) $x \to_n y$.

(ii) For every random variable w on Ω there is a random variable z on Γ such that $(x, w) \equiv_n (y, z)$.

(iii) For every infinite sequence of sets $A_1, A_2, \ldots$ in $\mathcal{F}_1$ there is an infinite sequence of sets $B_1, B_2, \ldots$ in $\mathcal{G}_1$ such that

$$(x, \mathbb{I}_{A_1}, \mathbb{I}_{A_2}, \ldots) \equiv_n (y, \mathbb{I}_{B_1}, \mathbb{I}_{B_2}, \ldots).$$

(iv) For every Polish space M and M-valued random variable w on Ω there is an M-valued random variable z on Γ such that $(x, w) \to_n (y, z)$.

PROOF. Condition (i) is just the special case of (iv) where M is k-dimensional Euclidean space, so it is trivial that (iv) implies (i) and (i) implies (ii).

To see that (iii) implies (iv), let $D = \{d_1, d_2, \ldots\}$ be a countable dense subset of M. Given an M-valued random variable w on Ω, we may take a sequence $(w_1, w_2, \ldots)$ of random variables with values in D which converges to w in probability. Applying (iii) to the doubly indexed sequence of sets $A_{ij} = w_i^{-1}\{d_j\}$, we get a sequence $(z_1, z_2, \ldots)$ such that

$$(x, w_1, w_2, \ldots) \equiv_n (y, z_1, z_2, \ldots).$$

Basic Fact 1C.4 (f) from Chapter 1 tells us that the sequence $(z_1, z_2, \ldots)$ converges in probability to an M-valued random variable z on Γ, and that

$$(x, w_1, w_2, \ldots, w) \equiv_n (y, z_1, z_2, \ldots, z).$$

Therefore we have $(x, w) \equiv_n (y, z)$ as required.

The proof that (ii) implies (iii) uses the Cantor ternary set on the real line. Let $2^{\mathbb{N}}$ be the Polish space of sequences of 0's and 2's with a metric for the product topology. Let C, the Cantor ternary set, be the set of all real numbers $r \in [0, 1]$ such that the decimal representation of r is a sequence of 0's and 2's. Then C is a compact subset of $\mathbb{R}$ and the natural mapping $f : 2^{\mathbb{N}} \to C$ is a homeomorphism.

We can identify a sequence of sets $A = (A_1, A_2 \ldots)$ in $\mathcal{F}_1$ with a random variable on Ω taking values in $2^{\mathbb{N}}$. Then the function $w = f \circ A$ is a real valued random variable on Ω. By (ii), there is a random variable z on Γ such that $(x. w) \equiv_n (y, z)$. Then z takes values in the Cantor set C, and applying f^{-1} we obtain a sequence $B = (B_1, B_2, \ldots)$ of sets in $\mathcal{G}_1$ such that $z = f \circ B$. Since f is a homeomorphism it follows that

$$(x, \mathbb{I}_{A_1}, \mathbb{I}_{A_2}, \ldots) \equiv_n (y, \mathbb{I}_{B_1}, \mathbb{I}_{B_2}, \ldots).$$

and (iii) is proved. $\dashv$

Applying the above theorem in both directions, we immediately get corresponding characterizations of the equivalence relations $\leftrightarrow_n$.

4C. Game equivalence

There is one more equivalence relation that we want to discuss. It is the strongest among all the ones we have introduced, and with it we can easily solve the question posed at the end of last chapter. To motivate this new equivalence relation, we will mention in passing a whole sequence of intermediate equivalence relations which correspond to games where the players choose random variables.

As usual, we start with a pair of stochastic processes x and y on adapted spaces Ω and Γ. Using condition (ii) in Theorem 4B.2, the equivalence relation $x \leftrightarrow_\infty y$ may be expressed in terms of a game $G_1(x, y)$, or G_1 for short, between two players. The game G_1 has just one pair of moves. The first player, named Spoiler, chooses a random variable over one of the adapted spaces, and then the second player, Duplicator, chooses a random variable over the other adapted space. In this way a random variable x_1 over Ω and a random variable y_1 over Γ are chosen. Duplicator wins this play of the game $G_1(x, y)$ if

$$(x, x_1) \equiv (y, y_1).$$

We see at once from the definitions that

PROPOSITION 4C.1. $x \leftrightarrow_\infty y$ *if and only if Duplicator has a winning strategy for* $G_1(x, y)$. $\dashv$

We can now get a new sequence of stronger and stronger equivalence relations by considering longer games. In the game G_m, the two players alternate moves m times with Spoiler moving first. As before, in each move pair Spoiler chooses a random variable from one of the adapted spaces and then Duplicator chooses a random variable from the other. In this way random variables $x_1, \ldots, x_m$ over Ω and $y_1, \ldots, y_m$ over Γ are chosen. Spoiler is allowed to choose an x_j over Ω for some moves and a y_j over Γ for other moves. Duplicator wins this play of the game $G_m(x, y)$ if

$$(x, x_1, \ldots, x_m) \equiv (y, y_1, \ldots, y_m).$$

The reader may notice the similarity between the game G_m and the Ehrenfeucht-Fraisse game in classical model theory. Let us write $x \equiv_{G_m} y$ if Duplicator has a winning strategy for the game $G_m(x, y)$. Proposition 4C.1 says that

$$x \leftrightarrow_\infty y \text{ if and only if } x \equiv_{G_1} y.$$

The reader can check that $x \equiv_{G_m} y$ is an equivalence relation. Moreover, if $x \equiv_{G_{m+1}} y$ then $x \equiv_{G_m} y$, because the beginning of a winning strategy for $G_{m+1}(x, y)$ will be a winning strategy for $G_m(x, y)$. Thus we have another infinite sequence of stronger and stronger equivalence relations between stochastic processes. Once more one can ask whether these equivalence relations are different from each other, and whether they have the amalgamation property.

We now come to the game we are really interested in, which was introduced in Fajardo [1990a]. The game G is like G_m, but Spoiler and Duplicator take turns ω times with Spoiler moving first. Infinite sequences of random variables $x_1, x_2, \ldots$ over Ω and $y_1, y_2, \ldots$ over Γ are chosen, and Duplicator wins this play of the game $G(x, y)$ if

$$(x, x_1, x_2, \ldots) \equiv (y, y_1, y_2, \ldots).$$

DEFNITION 4C.2. *We say that x, y are* **game equivalent**, *in symbols* $x \equiv_G y$, *if Duplicator has a winning strategy for the game $G(x, y)$.*

As in the case of the weaker relations introduced earlier in this chapter, it is easy to see that if x and y are processes on saturated adapted spaces, then $x \equiv_G y$ if and only if $x \equiv y$.

The next theorem, from Fajardo [1990a], gives the solution of Problem 3E.8 which was posed at the end of the preceding chapter. The game G was built precisely to solve this problem.

THEOREM 4C.3. *Let x and x' be stochastic processes on adapted spaces Ω and Γ respectively.*

(a) If $x \equiv_G x'$ then for every process y on Ω then there exists a process y' on Γ such that $(x, y) \equiv_G (x', y')$, and for every process y' on Γ then there exists a process y on Ω such that $(x, y) \equiv_G (x', y')$.

(b) (Amalgamation Property) Whenever $x \equiv_G x'$, y is a process on Ω, and z' is a process on Γ, there exists an adapted space Λ and processes x'', y'', z'' on Λ such that

$$(x, y) \equiv_G (x'', y'') \text{ and } (x', z') \equiv_G (x'', z'').$$

(c) The relation $\equiv_G$ is an equivalence relation on stochastic processes.

(d) The game equivalence relation $\equiv_G$ is strictly between the adapted equivalence relation and the isomorphism relation, that is,

$$\equiv \; < \; \equiv_G \; < \; \simeq.$$

PROOF. (a) Suppose $x \equiv_G x'$. When Spoiler chooses a process y on Ω for his first move, the winning strategy for Duplicator in the game $G(x, x')$ will produce the required process y'. The case where Spoiler chooses a process on Γ is similar.

(b) follows from (a) by taking $\Lambda = \Gamma$, $x'' = x'$, $y'' = y'$, and $z'' = z'$.

We leave the proof of (c) as an exercise. Note that to prove transitivity, one must produce a winning Duplicator strategy for the game $G(x, z)$ when strategies for the games $G(x, y)$ and $G(y, z)$ are given.

(d) Example 4A.6 shows that $\equiv\, <\, \equiv_G$. To prove that $\equiv_G\, <\, \simeq$ one can use the fact that there exist two hyperfinite adapted spaces Ω, Γ with different external cardinalities. Now let x and x' be the processes on these spaces with constant value 0. We leave the proof that $x \equiv_G x'$ but not $x \simeq x'$ as an exercise for the reader.   ⊣

Consider a pair of stochastic processes x and y. For each $m \in \mathbb{N}$, the game $G_m(x, y)$ with m move pairs is obviously determined, that is, at least one of the players Duplicator and Spoiler has a winning strategy. At this point it is natural to ask whether the game $G(x, y)$, which has infinitely many moves. is necessarily determined. If not, then the relation $x \sim y$, which holds if Spoiler does not have a winning strategy for the game $G(x, y)$, would be strictly weaker than the relation $x \equiv_G y$. But the next theorem shows that the game $G(x, y)$ is determined, so the relation $x \sim y$ is the same as $x \equiv_G y$.

THEOREM 4C.4. *For each pair of stochastic processes x, y, exactly one of the players Duplicator and Spoiler has a winning strategy for the game $G(x, y)$.*

PROOF. It is clear that both players cannot have winning strategies. Let $x \sim y$ mean that Spoiler does not have a winning strategy for the game between x and y. Assume $x \sim y$. We construct a winning strategy for Duplicator.

Claim: For every x_1 there exists y_1 with $(x, x_1) \sim (y, y_1)$, and for every y_1 there exists x_1 such that $(x, x_1) \sim (y, y_1)$.

The claim is true because otherwise Spoiler would have a winning strategy for the game G between x and y; for instance, if there exists y_1 such that for every x_1 not $(x, x_1) \sim (y, y_1)$, Spoiler first plays x_1, and after Duplicator replies y_1, Spoiler follows his winning strategy for G between (x, x_1) and (y, y_1).

We now construct a strategy S for Duplicator. If Spoiler plays x_1, Duplicator responds with the y_1 which exists by the claim. Similarly, if Spoiler plays y_1 then Duplicator responds with x_1 as in the claim. In either case, we have $(x, x_1) \sim (y, y_1)$. We now repeat this process ω times to define the strategy S. For example, if Spoiler plays x_{k+1}, we apply the claim starting with

$$(x, x_1, \ldots, x_k) \sim (y, y_1, \ldots, y_k)$$

to get a response y_{k+1} for Duplicator so that

$$(x, x_1, \ldots, x_{k+1}) \sim (y, y_1, \ldots, y_{k+1}).$$

No matter how Spoiler moves, the strategy S for Duplicator will produce a play of the game such that

$$(\forall n \in \mathbb{N})(x, x_1, \ldots, x_n) \sim (y, y_1, \ldots, y_n).$$

It follows that

$$(\forall n \in \mathbb{N})(x, x_1, \ldots, x_n) \equiv (y, y_1, \ldots, y_n),$$

because if not $u \equiv v$ then every strategy for Spoiler will win the game G between u and v. Then by Basic Fact 1C.4 (d) we have

$$(x, x_1, x_2, \ldots) \equiv (y, y_1, y_2, \ldots),$$

so S is a winning strategy for Duplicator.   ⊣

We conclude this section with another problem that is still open.

PROBLEM 4C.5. *Is there an equivalence relation strictly between* $\equiv$ *and* $\equiv_G$ *that satisfies the amalgamation property?*

4D. Examples and counterexamples

The reader already may be tired of so many equivalence relations. Perhaps we can summarize what we have done so far in this chapter with the following problem:

Sort out this mess.

And another question:

Which of the equivalence relations in this chapter preserve which properties listed in Exercise 1D.8 (d)?

Don't be discouraged. Many of the questions spread throughout this chapter will be answered in this section. The answers come from the papers Peña [1993] and Fajardo and Peña [1997]. Here is a chart which visually represents the status of the equivalence relations studied so far.

Chart I

$$\leftrightarrow_0 \quad \leq \cdots \leq \quad \leftrightarrow_\infty \quad = \quad \equiv_{G_1} \quad \leq \cdots \leq \quad \equiv_G \quad < \quad \simeq$$

$$\text{VI} \qquad\qquad\qquad \text{VI}$$

$$\equiv \quad < \quad \Leftrightarrow_0 \quad \leq \cdots \leq \quad \Leftrightarrow_\infty$$

In this chart, = means that the two relations are equivalent, $\leq$ means that the second relation implies the first, and < means that the second relation implies the first and there is an example showing that the two relations are not equivalent. $\simeq$ is the relation of being isomorphic. We will update this chart at the end of the section.

PROPOSITION 4D.1. $\equiv_{G_1} < \equiv_{G_2}$. *That is, there is a pair of stochastic processes* x, y *such that* $x \equiv_{G_1} y$ *but not* $x \equiv_{G_2} y$.

PROOF. Consider the following probability spaces Ω and Γ.

$$\Omega = ([0,1], \mathcal{B}([0,1]), \lambda)$$

is the probability space on the unit interval where $\mathcal{B}([0,1])$ is the Borel σ-algebra and λ is Lebesgue measure.

$$\Gamma = (\mathbb{T}, L(\mathbb{T}), L(\mu))$$

is the probability space where $\mathbb{T}$ is our hyperfinite time line, $L(\mathbb{T})$ is the Loeb σ-algebra generated by the internal subsets of $\mathbb{T}$, and $L(\mu)$ is the Loeb measure obtained from the internal counting measure μ on $\mathbb{T}$. Make Ω into an adapted space by giving it the constant filtration $(\mathcal{F}_t)$ where $\mathcal{F}_t = \mathcal{B}[0,1]$ for all t. Do the same with the space Γ, giving it the constant filtration $(\mathcal{G}_t)$ where $\mathcal{G}_t = L(\mathbb{T})$ for all t.

Since both of these adapted spaces are universal for random variables, we have $(\Omega, 0) \hookrightarrow_0 (\Gamma, 0)$, where 0 is the constant random variable zero. Moreover, and this is crucial here, since the filtrations are constant the conditional expectation is irrelevant. Therefore

$$(\Omega, 0) \hookrightarrow_\infty (\Gamma, 0),$$

so

$$(\Omega, 0) \equiv_{G_1} (\Gamma, 0).$$

We will show that it is not the case that

$$(\Omega, 0) \equiv_{G_2} (\Gamma, 0).$$

Let x be the identity function on $[0,1]$. Then x is Lebesgue measurable and hence is a random variable on Ω. Define $w : \mathbb{T} \to [0,1]$ by

$$w(t) = \begin{cases} 2^\circ t & \text{if } 0 \le t \le \frac{1}{2}, \\ 2^\circ t - 1 & \text{if } \frac{1}{2} < t \le 1. \end{cases}$$

Then w is Loeb measurable and is a random variable on Γ. By computing the distribution functions of x and w we see that

$$x \equiv_0 w.$$

Now consider any random variable y on Γ such that

$$(0, x) \equiv (0, y).$$

Since $x \equiv_0 w$ and at the same time $x \equiv_0 y$, we have $y \equiv_0 w$. We can now apply the homogeneity theorem to conclude that there is an automorphism $h : \mathbb{T} \to \mathbb{T}$ such that

$$y = w \circ h \quad a.s.$$

Let $B = h([0, 1/2])$. We claim that there is no $A \in \mathcal{B}([0,1])$ such that

$$(0, x, \mathbb{I}_A) \equiv_0 (0, y, \mathbb{I}_B).$$

To see this, suppose there were such an A. Then

$$\lambda(A) = L(\mu)(B) = 1/2.$$

But for each $a \in [0,1]$ we have:

$$\lambda(\{\omega \in A : x(\omega) \le a\}) = L(\mu)(\{t \in B : y(t) \le a\}) =$$
$$L(\mu)(\{t \in [0, 1/2] : w(t) \le a\}) = a/2.$$

It follows that

$$\lambda(A \cap C) = \lambda(C)/2$$

for each Borel set $C \in \mathcal{B}([0,1])$. Putting $A = C$ we have $\lambda(A) = 0$. This is a contradiction, and the claim is proved.

We can now conclude that $(\Omega, 0) \equiv_{G_2} (\Gamma, 0)$ fails. For Spoiler can choose x, Duplicator must reply with a y on Γ such that $(0, x) \equiv (0, y)$, then Spoiler can choose the set B defined from y as indicated above, and Duplicator has no winning reply.   ⊣

Now we are ready to tackle Question 4A.7, which asks whether or not $\leftrightarrow_0$ and $\Leftrightarrow_0$ are the same. One would be tempted to guess that they are the same since an arbitrary random variable can be approximated by a sequence of simple random variables. But the problem is that the relation $\Gamma \Leftrightarrow_0 \Omega$ does not guarantee that the two spaces have the same limits of simple random variables. Peña proved that the relations $\leftrightarrow_0$ and $\Leftrightarrow_0$ are different. Let us see what he did in Peña [1993].

The following lemma, which we leave as an exercise, is a direct consequence of Lemma 4A.3.

LEMMA 4D.2. *Let x and y be square integrable random variables on adapted spaces $(\Omega, \mathcal{F}_t, P)$ and $(\Gamma, \mathcal{G}_t, Q)$ respectively. If $x \leftrightarrow_0 y$ then for every $s \in [0, 1]$ we have*

$$(x, E[x|\mathcal{F}_s]) \leftrightarrow_0 (yE[y|\mathcal{G}_s]).$$ ⊣

The example in the next lemma gives us a pair of stochastic processes which are $\Leftrightarrow_0$ equivalent but not too much alike. and will be used several times.

LEMMA 4D.3. *Let C and D be the sets*

$$C = \{(r, s) \in [0. 1]^2 : s \leq r\},$$

$$D = \{(r, s) \in [0, 1]^2 : s \leq 2r \leq 1 \text{ or } s \leq 2r - 1\}.$$

Let Ω, Γ be the adapted spaces

$$\Omega = ([0, 1]^2, \mathcal{F}_t, P),$$

$$\Gamma = ([0, 1]^2, \mathcal{G}_t, P)$$

where the filtrations are defined by

$$\mathcal{F}_t = \mathcal{G}_t = \sigma(\mathcal{B}([0, 1]) \times \{[0, 1]\}) \text{ for } t \in [0, \frac{1}{2}),$$

$$\mathcal{F}_t = \sigma(\mathcal{F}_0, C) \text{ for } t \in [\frac{1}{2}, 1],$$

$$\mathcal{G}_t = \sigma(\mathcal{F}_0, D) \text{ for } t \in [\frac{1}{2}, 1].$$

Then

$$(\Omega, \mathbb{I}_C) \Leftrightarrow_0 (\Gamma, \mathbb{I}_D).$$

PROOF. The sets C and D are shown in Figure 1.
We leave the following three facts as exercises for the reader.
(1) $\mathbb{I}_C \equiv_0 \mathbb{I}_D$.
(2) Given disjoint sets $A_1, \ldots, A_n \in \mathcal{F}_0$, there exist disjoint sets $B_1, \ldots, B_n \in \mathcal{F}_0$ such that

$$P[A_i] = P[B_i] \text{ and } P[C \cap A_i] = P[D \cap B_i] \text{ for } i = 1, 2, \ldots, n.$$

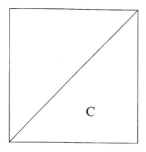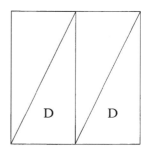

FIGURE 1

(3) Given disjoint sets $B_1, \ldots, B_n \in \mathcal{F}_0$, there exist disjoint sets $A_1, \ldots, A_n \in \mathcal{F}_0$ such that

$$P[A_i] = P[B_i] \text{ and } P[C \cap A_i] = P[D \cap B_i] \text{ for } i = 1, 2, \ldots, n.$$

Hint for (3): First prove that for every open set $U \subseteq [0, 1/2)$ there is an open set $V \subseteq 2U$ such that

$$P[U] = P[V] \text{ and } P[C \cap (V \times [0, 1])] = P[D \cap (U \times [0, 1])].$$

We now prove that

$$(\Omega, \mathbb{I}_C) \Rightarrow_0 (\Gamma, \mathbb{I}_D).$$

Let $F_1, \ldots, F_m \in \mathcal{F}_0$ and $H_1, \ldots, H_n \in \mathcal{F}_1$ be given. We must find sets $G_i \in \mathcal{G}_0$ and $K_j \in \mathcal{G}_1$ such that

(4) $(\mathbb{I}_C, \mathbb{I}_{F_1}, \ldots, \mathbb{I}_{F_m}, \mathbb{I}_{H_1}, \ldots, \mathbb{I}_{H_n}) \equiv_0 (\mathbb{I}_D, \mathbb{I}_{G_1}, \ldots, \mathbb{I}_{G_m}, \mathbb{I}_{K_1}, \ldots, \mathbb{I}_{K_n}).$

There is a finite list of disjoint sets $A_1, \ldots, A_p \in \mathcal{F}_0$ such that each F_i is a finite union of A_k's, and each H_j is a finite union of sets of the form $A_k \cap C$ or $A_k \setminus C$. By (2) we can find disjoint sets $B_1, \ldots, B_p \in \mathcal{F}_0$ such that the conclusion of (2) holds. Let G_i and K_j be the sets built from the B_k's and D in the same way that the given sets F_i and H_j are built from the A_k's and C. Then

$$G_1, \ldots, G_m \in \mathcal{G}_0, \quad K_1, \ldots, K_n \in \mathcal{G}_1,$$

and (4) holds. This proves

$$(\Omega, \mathbb{I}_C) \Rightarrow_0 (\Gamma, \mathbb{I}_D).$$

The other direction is proved by a similar argument using (3) instead of (2). ⊣

The next example, from [P], is the negative answer we promised for Hoover's Question 4A.7.

EXAMPLE 4D.4. *There are stochastic processes x, y such that $x \Leftrightarrow_0 y$ but not $x \leftrightarrow_0 y$. Thus $\Leftrightarrow_0 < \leftrightarrow_0$.*

PROOF. For this example we will build upon Lemma 4D.3 and take $\mathbb{I}_C$ as x and $\mathbb{I}_D$ as y. By the lemma, $x \Leftrightarrow_0 y$. Now let $w = E[x|\mathcal{F}_0]$, $z = E[y|\mathcal{G}_0]$.

It is clear that $w(r, s) = r$ and

$$z(r, s) = \begin{cases} 2r & \text{if } 0 \le r \le \frac{1}{2} \\ 2r - 1 & \text{if } \frac{1}{2} < r \le 1. \end{cases}$$

As in the proof of Proposition 4D.1, if we take

$$B = [0, \frac{1}{2}) \times [0, 1],$$

there does not exist $A \in \mathcal{F}_0$ such that $(w, \mathbb{I}_A) \equiv_0 (z, \mathbb{I}_B)$. Thus $w \Leftrightarrow_0 z$ fails, hence $w \hookleftarrow_0 z$ fails, and so

$$(x, E[x|\mathcal{F}_0]) \hookleftarrow_0 (y, E[y|\mathcal{G}_0]) \text{ fails.}$$

Then by Lemma 4D.2 we cannot have $x \hookleftarrow_0 y$.   ⊣

Notice that the preceding example also shows that Lemma 4D.2 does not hold with $\Leftrightarrow_0$ in place of $\hookleftarrow_0$. In fact, $x \Leftrightarrow_0 y$ holds but

$$(x, E[x|\mathcal{F}_0]) \Leftrightarrow_0 (y, E[y|\mathcal{G}_0]) \text{ fails.}$$

This same example can be modified in the obvious way in order to show that none of the relations $\Leftrightarrow_n$, with $n = 1, 2, \ldots, \infty$, implies $\hookleftarrow_0$. For details see [P].

The following result of Peña is somewhat surprising in view of the preceding example. It shows that Hoover's question has a "Yes" answer for adapted spaces with constant filtrations. In other words, the answer to the analog of Hoover's question for plain probability spaces is "Yes."

PROPOSITION 4D.5. *Let $\Omega = (\Omega, \mathcal{F}, P)$ and $\Gamma = (\Gamma, \mathcal{G}, Q)$ be probability spaces with the following property:*

For every $n \in \mathbb{N}$ and $A_1, A_2, \ldots, A_n \in \mathcal{F}$ there exist $B_1, B_2, \ldots, B_n \in \mathcal{G}$ such that

$$(\mathbb{I}_{A_1}, \mathbb{I}_{A_2}, \ldots, \mathbb{I}_{A_n}) \equiv_0 (\mathbb{I}_{B_1}, \mathbb{I}_{B_2}, \ldots, \mathbb{I}_{B_n}).$$

Then given a random variable x on Ω there is a random variable y on Γ such that $x \equiv_0 y$.

PROOF. It suffices to show that given any countable partition $A_1, A_2, \ldots$ of Ω with $A_k \in \mathcal{F}$ and $P[A_k] > 0$ for each $k \in \mathbb{N}$, there is a partition $B_1, B_2, \ldots$ of Γ such that $B_k \in \mathcal{G}$ and $P[A_k] = Q[B_k]$ for each $k \in \mathbb{N}$. By hypothesis, for each $n \in \mathbb{N}$ there is a sequence $B_{n1}, B_{n2}, \ldots, B_{nn}$ of disjoint sets $B_{nk} \in \mathcal{G}$ with $P[A_k] = Q[B_{nk}]$ for each $k \le n$. The difficulty is that the sets B_{nk} may change as n increases. For instance, they will have to change if the complement of $B_{n1} \cup \ldots \cup B_{nn}$ contains only atoms of measure greater than $P[A_{n+1}]$. The main idea is to use the fact that for each atom $C \in \mathcal{G}$, there are only finitely many k such that $P[A_k] \ge Q[C]$ because $Q[C] > 0$. Hence there is an infinite subset $M \subseteq \mathbb{N}$ such that for each $k \in \mathbb{N}$, the sets $B_{mk}, k \le m \in M$, agree on each atom in $\mathcal{G}$. One can now easily adjust the sets B_{mk} on the atomless part of $\mathcal{G}$ to get a sequence B_k as required. For the details see [P].   ⊣

COROLLARY 4D.6. *If Ω and Γ are adapted spaces with constant filtrations and $\Omega \Rightarrow_0 \Gamma$, then $\Omega \to_0 \Gamma$.*   ⊣

We can now easily construct an example that allows us to show several things at the same time:

$$\Leftrightarrow_0 \; < \; \Leftrightarrow_1, \qquad \leftrightarrow_0 \; < \; \leftrightarrow_1,$$

and $\leftrightarrow_0$ does not imply $\Leftrightarrow_1$.

EXAMPLE 4D.7. *There are two stochastic processes x, y such that $x \leftrightarrow_0 y$ but not $x \Leftrightarrow_1 y$.*

PROOF. The idea is that between two zero processes, $\leftrightarrow_0$ does not allow us to distinguish between a constant filtration and a filtration which changes over time, but $\Leftrightarrow_1$ does. Let $\Omega = ([0, 1], \mathcal{F}_t, P)$ with the constant filtration $\mathcal{F}_t = \mathcal{B}[0, 1]$, and let $\Gamma = (\Gamma, \mathcal{G}_t, Q)$ be a hyperfinite adapted space. Since both Ω and Γ are atomless, one can easily check that $(\Omega, 0) \leftrightarrow_0 (\Gamma, 0)$. Now take a set B in Γ adapted at time t and not adapted at time s, with $s < t$ (i.e., $B \in \mathcal{G}_t$ but $B \notin \mathcal{G}_s$). It is impossible to find a set A in $\mathcal{F}_1$ with $(\Omega, \mathbb{I}_A) \equiv_1 (\Gamma, \mathbb{I}_B)$, because $A \in \mathcal{F}_s$, so $E[\mathbb{I}_A | \mathcal{F}_t] = \mathbb{I}_A$, but $E[\mathbb{I}_B | \mathcal{G}_s] \neq \mathbb{I}_B$. This shows that $(\Omega, 0) \Leftrightarrow_1 (\Gamma, 0)$ fails. ⊣

Here is an example from [P] (which also appears in Fajardo and Peña [1997]), that gives a negative solution to the amalgamation Problem 4B.1 (b) for the relation $\Leftrightarrow_0$.

PROPOSITION 4D.8. *The relation $\Leftrightarrow_0$ does not have the amalgamation property.*

PROOF. Consider the processes from Example 4D.4, and recall that we have $x \Leftrightarrow_0 y$ where $x = \mathbb{I}_C$ and $y = \mathbb{I}_D$, but not $w \Leftrightarrow_0 z$ where $w = E[x | \mathcal{F}_0]$ and $z = E[y | \mathcal{G}_0]$. Suppose that there exists an adapted space $\Lambda = (\Lambda, \mathcal{H}_t, R)$ and processes s, t and u on Λ such that

$$(8) \qquad (s, t) \Leftrightarrow_0 (x, w) \text{ and } (s, u) \Leftrightarrow_0 (y, z).$$

We shall show that under these conditions we get $u = t = E[s | \mathcal{H}_0]$, so that $(x, w) \Leftrightarrow_0 (y, z)$ and consequently $w \Leftrightarrow_0 z$, which is a contradiction.

It follows from our assumption (8) that $(s, t) \equiv_0 (x, w)$, $s \Leftrightarrow_0 x$, and t is $\mathcal{H}_0$-measurable. Since $s \Leftrightarrow_0 x$, there exists an $\mathcal{F}_0$-measurable v on Ω such that $(s, E[s | \mathcal{H}_0]) \equiv_0 (x, v)$. Since $w = E[x | \mathcal{F}_0]$, we have $(s, t) \equiv_0 (x, E[x | \mathcal{F}_0])$. Thus by Lemma 4A.3, $t = E[s | \mathcal{H}_0]$ a.s.

A similar argument shows that $u = E[s | \mathcal{H}_0]$ a.s., and this completes the proof. ⊣

In Problem 4B.1 (a) we asked whether or not the relations $\leftrightarrow_n, n = 1, 2, \dots$ are different. We are now going to see what happens.

THEOREM 4D.9. *All the equivalence relations $\leftrightarrow_1, \leftrightarrow_2, \dots, \leftrightarrow_n, \dots, \leftrightarrow_\infty$ are the same.*

PROOF. We must show that $\leftrightarrow_\infty \leq \leftrightarrow_1$. Suppose that $x \leftrightarrow_1 y$, and that the process x is defined on Ω and y is defined on Γ. Let u be a random variable on Ω. By Theorem 4B.2, to prove $x \leftrightarrow_\infty y$ it is sufficient to find a random variable v on Γ such that $(x, u) \equiv (y, v)$.

We first give the proof in the case that x and y are cadlag processes. For any random variables u and v, (x, u) and (y, v) are also cadlag. By Corollary 3D.5

there is a countable set $F = \{f_n : n \in \mathbb{N}\}$ of adapted functions such that for each random variable v,

(9) $(x, u) \equiv (y, v)$ if and only if $E[f_n(x, u)] = E[f_n(y, v)]$ for all n.

Since $x \leftrightarrow_1 y$, by Theorem 4B.2 there is a random variable v and a sequence of random variables $(z_n : n \in \mathbb{N})$ on Γ such that

(10) $(x, u, (f_n(x, u))_{n \in \mathbb{N}}) \equiv_1 (y, v, (z_n)_{n \in \mathbb{N}}).$

Now by induction on the complexity of f_n we will show that

(11) $z_n = f_n(y, v) \ a.s.$

The Basis and Composition Steps in the induction are easy and are left as exercises.

Conditional Expectation Step: Let $f_n = E[f_i|\mathcal{F}_t]$. Then

$$(x, u, f_i(x, u), f_n(x, u)) \equiv_1 (y, v, z_i, z_n).$$

We have $f_n(x, u) = E[f_i(x, u)|\mathcal{F}_t]$. By the definition of $\equiv_1$, it follows that $z_n = E[z_i|\mathcal{G}_t]$. By induction hypothesis, $z_i = f_i(y, v) \ a.s.$ Therefore

$$z_n = E[f_i(y, v)|\mathcal{G}_t] = f_n(y, v) \ a.s.$$

By (10) and (11) we obtain

$$(x, u, (f_n(x, u))_{n \in \mathbb{N}}) \equiv_1 (y, v, (f_n(y, v))_{n \in \mathbb{N}}).$$

Therefore $E[f_n(x, u)] = E[f_n(y, v)]$ for all n, and by (9) we have $(x, u) \equiv (y, v)$.

The general case of the theorem can be obtained from the cadlag case by using Lemma 1E.3 (b).   ⊣

We can summarize the results as follows:

Summary

$$\equiv \ < \ \leftrightarrow_0 \ < \ \leftrightarrow_1 \ \leq \ \leftrightarrow_2 \ \leq \ \cdots \ \leq \ \leftrightarrow_\infty \ < \ \leftrightarrow_1$$
$$\leftrightarrow_0 \ < \ \leftrightarrow_0 \ < \ \leftrightarrow_1 \ = \ \leftrightarrow_2 \ = \ \cdots \ = \ \leftrightarrow_\infty$$
$$\leftrightarrow_\infty \ = \ \equiv_{G_1} \ < \ \equiv_{G_2} \ \leq \ \cdots \ \leq \ \equiv_G \ < \ \simeq$$

$\leftrightarrow_\infty$ does not imply $\leftrightarrow_0$.

$\leftrightarrow_0$ does not imply $\leftrightarrow_1$.

Let us bring our previous Chart I up to date in view of the results of this section.

Chart II

$$\leftrightarrow_0 \ < \ \leftrightarrow_1 = \ \cdots \ = \leftrightarrow_\infty \ = \ \equiv_{G_1} \ < \ \equiv_{G_2} \ \leq \cdots \leq \ \equiv_G \ < \ \simeq$$

$$\lor \qquad \lor \qquad\qquad \lor$$

$$\equiv \ < \ \leftrightarrow_0 < \leftrightarrow_1 \leq \ \cdots \ \leq \leftrightarrow_\infty$$

CHAPTER 5

DEFINABILITY IN ADAPTED SPACES

The Homogeneity Theorem from Chapter 2 showed us that in a hyperfinite adapted space two processes with the same adapted distribution are the same in a precise mathematical sense. They are automorphic, that is, there is an adapted isomorphism from the space to itself which carries one process to the other almost surely. The notion of automorphism used is very strong—one process is essentially a renaming of the other. What can be said of processes that live on possibly different adapted spaces and have the same adapted distribution? Do we have any result that resembles the strong characterization of adapted equivalence in a hyperfinite adapted space? There is an answer: they are isomorphic, but in a weaker sense. In Section 5A we will prove the Intrinsic Isomorphism Theorem, which makes this statement rigorous. It is due to Hoover and Keisler, but was first published in Section 3 of Fajardo [1987] with some minor generalizations. This result was used in Hoover [1992]. Section 5B is based on both of the papers Fajardo [1987] and Hoover [1992], and uses the notions introduced in Section 5A for a theory of definability in adapted spaces.

5A. The intrinsic filtration

In Chapter 1 we gave an informal argument explaining the need for stochastic processes on adapted probability spaces as mathematical structures which model information that evolves with time. Now we can ask a question that takes us further. *Given a stochastic process x on an adapted space Ω, what is the smallest filtration—i.e., the minimal amount of information—needed to capture the behavior of the process as it evolves over time?* The next definition gives a useful answer.

DEFINITION 5A.1. *Let Ω be an adapted space and x a stochastic process on Ω.*

*(a) The **intrinsic σ-algebra** of x on Ω, denoted $\mathcal{I}^x$, is the σ-algebra generated by the family of random variables $\{f(x) : f \in AF\}$ and the set $\mathcal{N}$ of null sets of Ω.*

*(b) The **intrinsic filtration** of x with respect to Ω is the filtration $(\mathcal{F}_t \cap \mathcal{I}^x)_{t \in [0,1]}$. This filtration is denoted by $(\mathcal{I}_t^x)$. The adapted space $(\Omega, \mathcal{I}_t^x, P)$ is denoted Ω^x.*

*(c) x is said to be **intrinsic** with respect to the adapted space $(\Omega, \mathcal{F}_t, P)$ if $\Omega^x = (\Omega, \mathcal{F}_t, P)$. Observe that every process x is intrinsic with respect to Ω^x.*

We also introduce the analogous notions for adapted functions of rank at most r (see Definition 1D.1).

Definition 5A.2. *Let AF^r be the set of all adapted functions of rank at most r.*

(a) The r-intrinsic σ-algebra of x on Ω, denoted $\mathcal{I}^x_r$, is the σ-algebra generated by the family of random variables $\{f(x) : f \in AF^r\}$ and $\mathcal{N}$.

(b) The r-intrinsic filtration is defined in the obvious way and is denoted by $(\mathcal{I}^x_{r,t})$. The adapted space $(\Omega, (\mathcal{I}^x_{r,t}), P)$ is denoted Ω^x_r

(c) x is said to be r-intrinsic with respect to $(\Omega, \mathcal{F}_t, P)$ if $\Omega^x_r = (\Omega, \mathcal{F}_t, P)$.

The above definitions can be extended in a natural way to the case where we have more than one stochastic process at a time, so we can talk about the intrinsic filtration $(\mathcal{I}^{x,x'}_t)$ associated with a pair of processes (x, x'), etc.

The following proposition contains the basic properties of the intrinsic filtrations. We will give proofs just for the intrinsic filtration. The r-case will always be proved in an identical way.

Proposition 5A.3. *Let x be a stochastic process on an adapted space Ω.*

(a) $(\Omega, x) \equiv (\Omega^x, x)$.

(b) For every r, $(\Omega, x) \equiv_r (\Omega^x_r, x)$.

(c) $\mathcal{I}^x_t = \sigma(\{f(x) : f \text{ is of the form } E[g|t]\}) \vee \mathcal{N}$.

(d) For each $r \geq 1$,
$\mathcal{I}^x_{r,t} = \sigma(\{f(x) : f \text{ is of the form } E[g|t], \text{ with } \mathrm{rank}(f) < r\}) \vee \mathcal{N}$.

Proof. To prove (a), we will prove the following stronger property. For every adapted function f.

$$f((\Omega, x)) = f((\Omega^x, x)) \ a.s.$$

The implications of this fact will be seen in a later chapter. As usual, the proof is by induction on f. The nontrivial step is the case $f = E[g|t]$. so suppose $g((\Omega, x)) = g((\Omega^x, x)) \ a.s.$ Then

$$f((\Omega, x)) = E[g(\Omega, x)|\mathcal{F}_t] = E[g(\Omega, x)|\mathcal{I}^x_t] = E[g(\Omega^x, x)|\mathcal{I}^x_t] = f((\Omega^x, x)) \ a.s.$$

The second equality is justified by the fact that by definition, $E[g(\Omega, x)|\mathcal{F}_t]$ is both $\mathcal{I}^x$ and $\mathcal{F}_t$-measurable.

We prove the interesting part of (c). Take a set A in $\mathcal{F}_t \cap \mathcal{I}^x$ and show that A belongs to

$$\sigma(\{f(x) : f \text{ is of the form } E[g|t]\} \vee \mathcal{N}.$$

Since $A \in \mathcal{I}^x$, A is of the form $f(x)^{-1}(C)$ for some adapted function f and some Borel set C. Now consider the adapted function $g = E[f|t]$, and verify that $A = f(x)^{-1}(C) = g(x)^{-1}(C) \ a.s.$ To see this, observe that if the a.s. equality is false, the set

$$B = \{\omega \in \Omega : E[f(x)|\mathcal{F}_t](\omega) \in C, \ f(x)(\omega) \in C, \ E[f(x)|\mathcal{F}_t](\omega) \neq f(x)(\omega)\}$$

has positive measure. Notice that this set belongs to $\mathcal{F}_t$, so without loss of generality we can assume there is an interval (a, b) contained in C such that the set

$$D = \{\omega : E[f(x)|\mathcal{F}_t] \geq b \text{ and } f(x) \in (a, b)\}$$

is in $\mathcal{F}_t$ and has positive probability. Thus $\int_D E[f(x)|\mathcal{F}_t]dP$ is different from $\int_D f(x)dP$, and by the definition of conditional expectation this is a contradiction.

Parts (b) and (d) are left as exercises.   ⊣

If we accept that adapted equivalence captures the notion of two processes being "alike", then part (a) of the preceding proposition tells us that we can safely restrict ourselves to the study of intrinsic filtrations. Later on we will see some consequences of this remark.

Now we can introduce a notion of isomorphism between adapted spaces and processes that goes with the adapted equivalence relation. The idea is to take a measure algebra isomorphism which is a mapping between equivalence classes of sets modulo the null sets, and add the additional condition that the mapping respects the filtration. Notice that we mix, and slightly change, the notation and names from Fajardo [1987] and Hoover [1992].

DEFINITION 5A.4. *(a) Let Ω and Γ be adapted spaces. For each t, let $\mathcal{F}_t/\mathcal{N}$ and $\mathcal{G}_t/\mathcal{N}$ be the measure algebras obtained by identifying sets modulo the ideal $\mathcal{N}$ of null sets in the corresponding space (notice the abuse of notation). A bijection*

$$h : \mathcal{F}_1/\mathcal{N} \to \mathcal{G}_1/\mathcal{N}$$

*is a **filtration isomorphism** if whenever $F, F' \in \mathcal{F}_1$ and G, G' are representatives of the equivalence classes of $h(F/\mathcal{N})$ and $h(F'/\mathcal{N})$ respectively,*
(i) $F \subseteq F'$ a.s. if and only if $G \subseteq G'$ a.s.
(ii) $P[F] = Q[G]$.
(iii) For each t, $F \in \mathcal{F}_t$ if and only if $G \in \mathcal{G}_t$.

(b) If x and y are stochastic processes on Ω and Γ respectively, a filtration isomorphism from (Ω, x) to (Γ, y) is a filtration isomorphism h from Ω to Γ such that for all Borel sets $B \subseteq \mathbb{R}$ and times t,

$$h((x_t \in B)/\mathcal{N}) = (y_t \in B)/\mathcal{N}.$$

Usually, we will not bother to write the $\mathcal{N}$, and we will write equations like the one above as almost sure equalities,

$$h(x_t \in B) = (y_t \in B) \, a.s.$$

The next proposition shows that a filtration isomorphism h from Ω to Γ induces a mapping $\tilde{h}$ from stochastic processes on Ω to stochastic processes on Γ, provided that we identify any two versions of a process.

PROPOSITION 5A.5. *Let h be a filtration isomorphism from Ω to Γ. Then for each stochastic process x on Ω there is a stochastic process $y = \tilde{h}(x)$ on Γ such that:*
(a) h is a filtration isomorphism from (Ω, x) to (Γ, y).
(b) The process y is unique up to a version. That is, h is a filtration isomorphism from (Ω, x) to (Γ, z) if and only if z is a version of y.
(c) If x is a random variable, then so is y.

PROOF. Parts (b) and (c) are clear. We prove (a). In the case that x is a simple random variable on Ω of the form $\Sigma\, a_i \mathbb{I}_{A_i}$. take for y the simple random variable $y = \Sigma\, a_i \mathbb{I}_{h(A_i)}$ on Γ.

Suppose next that x is an arbitrary random variable on Ω. Choose a sequence (x_n) of simple random variables such that $x_n \to x$ a.s. Then for each n, there is a simple random variable y_n on Γ such that h is a filtration isomorphism from (Ω, x_n) to (Γ, y_n). It follows that y_n converges in probability to an almost surely unique random variable y on Γ, and h is a filtration isomorphism from (Ω, x) to (Γ, y).

Finally, suppose x is a stochastic process on Ω. By the preceding paragraph, for each t there is a random variable $y(t)$ on Γ such that $\tilde{h}(x(t)) = y(t)$. Combine the random variables $y(t)$ to form a function $y : \Gamma \times [0, 1] \to \mathbb{R}$. We must still show that y is measurable in the product space $\Gamma \times [0, 1]$. x is measurable in the product $\Omega \times [0, 1]$, so we may choose a sequence (x_n) of simple step functions such that $x_n \to x$ a.s. in $\Omega \times [0, 1]$. By the Fubini theorem, for all t in a set $U \subseteq [0, 1]$ of measure one, $x_n(t) \to x(t)$ a.s. in Ω. Using the preceding paragraph again, there is a sequence of simple step functions y_n on $\Gamma \times [0, 1]$ with h a filtration isomorphism from each step of x_n to the corresponding step of y_n. Then there is a measurable $z : \Gamma \times [0, 1] \to \mathbb{R}$ such that $y_n \to z$ in probability. Also, for each $t \in U$, $y_n(t) \to y(t)$ in probability. It follows that $y = z$ a.s. in $\Gamma \times [0, 1]$. This shows that y is measurable and hence y is a stochastic process on Γ.   ⊣

Let us explicitly write down the definition of $\tilde{h}$ from the last proposition.

DEFNITION 5A.6. *Let x be a stochastic process on Ω and let h be a filtration isomorphism from Ω to Γ. $\tilde{h}(x)$ is the stochastic process y on Γ such that h is a filtration isomorphism from (Ω, x) to (Γ, y).*

The next lemma says that the mapping $\tilde{h}$ preserves some basic properties of random variables. For more on this see [R].

LEMMA 5A.7. *Let x and y be $\mathcal{F}_1$-measurable random variables on an adapted space Ω and let h be a filtration isomorphism from Ω to another adapted space Γ.*
(a) If $E[x]$ exists then $E[x] = E[\tilde{h}(x)]$.
(b) For each time t, x is $\mathcal{F}_t$-measurable if and only if $\tilde{h}(x)$ is $\mathcal{G}_t$-measurable.
(c) $\tilde{h}(x \cdot y) = \tilde{h}(x) \cdot \tilde{h}(y)$.   ⊣

We now show that the mapping $\tilde{h}$ commutes with all adapted functions on a stochastic process. This is the final fact needed in order to characterize the adapted equivalence of a pair of stochastic processes by means of a filtration isomorphism.

LEMMA 5A.8. *Let h be a filtration isomorphism from Ω to Γ and let x be a stochastic process on Ω. Then for each adapted function $f \in AF$,*
(a) $\tilde{h}(f(x)) = f(\tilde{h}(x))$,
(b) $E[f(x)] = E[f(\tilde{h}(x))]$.

PROOF. The proof of (a) is by induction on AF. In order to carry out the induction we will prove a slightly stronger result, that the lemma holds for all adapted Borel functions f. As in most cases we limit ourselves to proving the conditional expectation step in the induction and leave the rest as exercises for the interested reader.

We do (a) first. Let f be of the form $E[g|t]$ and assume that the lemma holds for g. Let $y = \widetilde{h}(x)$. We have to verify that for each $r \in \mathbb{Q}$,

$$h(f(x) \geq r) = (f(y) \geq r).$$

Suppose this condition fails for some $r \in \mathbb{Q}$. Then if U and Z are $\mathcal{G}_t$-representatives of $h(f(x) \geq r)$ and $(f(y) \geq r)$ respectively, we can assume without loss of generality that $V = U - Z$ has positive measure. Since V is also in $\mathcal{G}_t$ and h is a filtration isomorphism, $V/\mathcal{N}$ has a preimage $W/\mathcal{N}$ in $\mathcal{F}_t/\mathcal{N}$. Observe that $V \subseteq U$, so $W \subseteq (f(x) \geq r)/\mathcal{N}$ and

$$\int_W g(x)dP = \int_W E[g(x)|\mathcal{F}_t]dP > rP[W].$$

On the other side the corresponding relation is

$$\int_V g(y)dQ = \int_V E[g(y)|\mathcal{G}_t]dQ < rQ[V].$$

But these two inequalities together with the induction hypothesis lead us to a contradiction which completes the proof of (a):

$$\int_V g(y)dQ < rQ[V] = rP[W] < \int_W g(x)dP.$$

The attentive reader will notice that this last statement is based on the facts that h is measure preserving, that $P[W] = Q[V]$, and that by the preceding lemma,

$$\int_A zdP = \int_{h(A)} \widetilde{h}(z)dQ$$

for any random variable z on Ω. Finally, letting A be Ω and using (a) we obtain (b).   ⊣

DEFINITION 5A.9. *A filtration isomorphism h from (Ω, x) to (Γ, y) is said to be an* **intrinsic isomorphism** *if the restriction of h to the intrinsic measure algebra $\mathcal{I}^x/\mathcal{N}$ is a filtration isomorphism from (Ω^x, x) to (Γ^y, y).*

In this chapter we have been assuming our processes to be real-valued. As usual, there is an obvious way of extending the definitions and results from this chapter to M-valued stochastic processes with M a Polish space. The reader should have no problem defining the notion of r-filtration isomorphism, and extending the definition of intrinsic isomorphism to the case where we have more than one process on each space.

Everything is in place now for the main result of this section, the Intrinsic Isomorphism Theorem.

THEOREM 5A.10. (*Intrinsic Isomorphism Theorem*). *Let x and y be stochastic processes on adapted spaces Ω and Γ respectively. Then $x \equiv y$ if and only if there exists a unique intrinsic isomorphism from (Ω^x, x) to (Γ^y, y).*

PROOF. From right to left the proof follows immediately from (b) in Lemma 5A.8. Now for the other direction. There is an obvious way of defining the isomorphism h. For each Borel set B and adapted function f, let

$$h(f(x) \in B/\mathcal{N}) = (f(y) \in B/\mathcal{N}).$$

The proof that h is an intrinsic isomorphism is as obvious as its definition, and uniqueness also follows from Lemma 5A.8.   ⊣

In Fajardo [1987] the reader can find natural extensions of this theorem to cases where we consider more than one stochastic process at a time in one given adapted space, and the analogous results where adapted equivalence $\equiv$ is replaced by $\equiv_r$. We will freely use such extensions of the theorem whenever we need them. For another proof see Hoover [1992].

There are several possible uses of this characterization theorem. It can be used for theoretical purposes (i.e., to continue building the theory) as in Fajardo [1990a] and Hoover [1992], or for applications to the theory of stochastic processes as indicated in Fajardo [1987]. We will turn to the applications in the following section. and leave further theoretical results to later chapters in this book.

5B. Intrinsic definability

The motivation for the material in this section was explained in Fajardo [1987], and was partially discussed in the final comments in Chapter 2. As usual, the main ideas were inspired by classical model theory. We start by asking for an adequate notion of definability for stochastic processes on adapted spaces. A natural and obvious first attempt is to call a process y definable from x if y is obtained by a direct application of a continuous function to x. This is very restrictive and of little interest for us within our framework. The following definition gives a broader alternative and is the starting point for the applications we are going to present.

DEFINITION 5B.1. *Let x and y be stochastic processes on the same adapted space Ω. We say that y is **definable** from x (or y is x-definable) if $\mathcal{I}^y \subseteq \mathcal{I}^x$. Similarly, we say that y is r-definable from x if $\mathcal{I}_r^y \subseteq \mathcal{I}_r^x$.*

A simple and quite important example helps to illustrate this concept. Consider an adapted cadlag process x and let B be a Borel subset of $\mathbb{R}$. Define the first entrance (or hitting) time of x in B by the formula

$$S_{x,B}(\omega) = \inf\{t \in [0, 1] : x(\omega, t) \in B\}.$$

It is well known that $S_{x,B}$ is a stopping time with respect to Ω. In fact, this result still holds for a more general class of processes x and sets B (see Dellacherie and Meyer [1978]).

PROPOSITION 5B.2. *For every adapted cadlag stochastic process x on an adapted space Ω, the stochastic process $S_{x,B}$ is 1-definable from x.*

PROOF. The argument is basically a repetition of the proof that hitting times are stopping times (see Dellacherie and Meyer [1978]), and we leave it as an exercise.   ⊣

The next theorem from Fajardo [1987] shows something interesting and intuitively clear. If we have two processes with the same adapted distribution, then whatever is definable from one must have a corresponding object which is definable from the other. This result allows us to give saturation-like arguments in adapted spaces that are neither hyperfinite nor saturated. As we will see later, this idea is closely related to the Come-back Problem discussed in Chapter 2.

THEOREM 5B.3. (*Definability Theorem*). *Let x and y be stochastic processes on adapted spaces Ω and Γ respectively. If $x \equiv y$ and z is a stochastic process on Ω which is definable with respect to x, then there exists a stochastic process w on Γ such that:*
(i) w is definable with respect to y, and
(ii) $(x, z) \equiv (y, w)$.

PROOF. Assume first that z is cadlag. Let h be the unique intrinsic isomorphism from Ω^x to Γ^y given by the Intrinsic Isomorphism Theorem.

For each $r \in [0, 1]$ let m_r be the least $s \in [0, 1]$ such that z_r is $\mathcal{I}^x \cap \mathcal{F}_s$-measurable. This s exists because the set

$$M_r = \{ s \in [0, 1] : z_r \text{ is } \mathcal{I}^x \cap \mathcal{F}_s\text{-measurable} \}$$

is nonempty, since z_r is $\mathcal{I}^z$-measurable and $\mathcal{I}^z \subseteq \mathcal{I}^x \subseteq \mathcal{F}_1$. We conclude that the set M_r has a greatest lower bound, and since the filtration $(\mathcal{F}_t)$ is right continuous, this bound is in fact the first element of M_r.

Fix a rational number r. Since z_r is $\mathcal{I}^x \cap \mathcal{F}_{m_r}$-measurable, by elementary measure theory we know that there exists a sequence (S_n) of $\mathcal{I}^x \cap \mathcal{F}_{m_r}$-measurable simple functions such that $\lim S_n = z_r$ a.s. Each S_n can be written as a finite linear combination $S_n = \sum s_i \mathbb{I}_{A_i}$ where $A_i = \{ S_n = s_i \} \subseteq \mathcal{I}^x \cap \mathcal{F}_{m_r}$ for each i, and the collection of A_i's partition Ω. For each A_i let $B_i \in \mathcal{I}^y \cap \mathcal{G}_{m_r}$ be such that $h(A_i/\mathcal{N}) = B_i/\mathcal{N}$. Then for each i, $P[A_i] = Q[B_i]$ and the B_i's partition Γ, modulo a null set. Now, for each n define the function $T_n = \sum s_i \mathbb{I}_{B_i}$, and define w_r as follows:

$$w_r(\gamma) = \lim_n T_n(\gamma) \text{ if the limit exists, } = 0 \text{ otherwise.}$$

We can now define w_t for arbitrary t by choosing a decreasing sequence of rational numbers (r_k) converging to t:

$$w_t(\gamma) = \lim_k w_{r_k}(\gamma) \text{ if the limit exists, } = 0 \text{ otherwise.}$$

We now claim that $(x, z) \equiv (y, w)$. It follows from the way w was defined from z through h that

$$\mathcal{I}^z \subseteq \mathcal{I}^x = \mathcal{I}^{x,z} \text{ and } \mathcal{I}^w \subseteq \mathcal{I}^y = \mathcal{I}^{y,w}.$$

The map h can now be regarded as a function from $\Omega^{x,z}$ to $\Gamma^{y,w}$. All we have to do now is to check that h is again an intrinsic isomorphism. But our construction of w from z makes this step simple. It boils down to verifying that in the above definition of w_r,

$$h(S_n \in B/\mathcal{N}) = (T_n \in B/\mathcal{N}).$$

The case where z is arbitrary follows from the cadlag case using Lemma 1E.3 (b).   ⊣

We can now give some applications of this theorem. The first application comes from Fajardo [1987] and is related to the well known operation of stopping a stochastic process at a stopping time. The stochastic process x stopped at the stopping time σ is denoted by x^σ.

THEOREM 5B.4. *Let x and y be cadlag processes on adapted spaces Ω and Γ respectively. If $x \equiv y$ and σ is a stopping time which is definable with respect to x, then there exists a stopping time θ which is definable with respect to y such that $(x, x^\sigma) \equiv (y, y^\theta)$.*

PROOF. The proof is a typical argument by induction on stopping times. We give a quick sketch:

(1) Suppose σ is a simple stopping time. Then there exists a simple y-definable stopping time θ such that $(x, \sigma) \equiv (y, \theta)$. The argument is identical to the one given in the proof of the above Definability Theorem.

(2) If σ and θ are as in (1), prove that $(x, x^\sigma) \equiv (y, y^\theta)$. The reason: x^σ is x-definable and y^θ is y-definable.

(3) Let σ be an arbitrary x-definable stopping time. Choose a decreasing sequence (σ_n) of simple x-definable stopping times converging to σ. The fact that the σ_n's can be chosen x-definable follows from the way these sequences are defined from x (see Dellacherie and Meyer [1978]). For each σ_n use (2) to obtain θ_n so that

$$(x, x^{\sigma_1}, \dots, x^{\sigma_n}, \dots) \equiv (y, y^{\theta_1}, \dots, y^{\theta_n}, \dots).$$

(4) Using the continuity properties of x and y and the fact that adapted equivalence is preserved under pointwise convergence we get

$$(x, \lim x^{\sigma_n}) \equiv (y, \lim y^{\theta_n}) \text{ and } \lim x^{\sigma_n} = x^\sigma.$$

Define θ as $\lim \theta_n$. This θ is a Γ-stopping time which has the desired properties.
                                                                          ⊣

REMARK 5B.5. *Each of the two preceding theorems also holds with $\equiv$ everywhere replaced by $\equiv_r$ and "definable" everywhere replaced by "r-definable".*

Here is an example, due to Hoover (see Fajardo [1987]), that shows that the definability hypothesis in the preceding theorem cannot be removed.

EXAMPLE 5B.6.   *Let $\Omega = \Gamma = [0, 1]$, for each t let $\mathcal{F}_t$ be the family of Borel subsets of $[0, 1]$, and let P be Lebesgue measure on $[0, 1]$. Define stochastic processes x and y as follows*:

$$x(\omega, t) = y(\omega, t) = 0 \text{ if } t < \omega$$
$$x(\omega, t) = 2 \text{ if } t = \omega$$
$$x(\omega, t) = 1 \text{ if } t > \omega$$
$$y(\omega, t) = 1 \text{ if } t \geq \omega$$

Clearly $x \equiv y$, but given the stopping time $\sigma(\omega) = \omega$ we cannot find a stopping time θ such that $(x, x^\sigma) \equiv (y, y^\theta)$.

A result showing that two stochastic processes are $\equiv_n$-equivalent will get better as n increases. We want to know which probabilistic properties which commonly arise in the theory of processes are preserved under the different $\equiv_n$'s. One important property is that of being a local martingale. Hoover [1984] proved that the local martingale property is preserved under $\equiv_1$. His proof in the general case used notions about stochastic processes which we do not want to introduce here. As an illustration, we will prove a particular case of Hoover's result which follows easily from the results presented in this chapter.

PROPOSITION 5B.7.   *Let x and y be continuous stochastic processes such that x is a local martingale and $x \equiv_1 y$. Then y is a local martingale.*

PROOF.   We use the following result from Durrett [1984], page 51: *Given a continuous local martingale x, the sequence $T_n = \inf\{t : |x_t| > n\}$ localizes x.* Each T_n is 1-definable from x by Proposition 5B.2. Then using the Definability Theorem for $\equiv_1$, we can find a sequence (S_n) of Γ-stopping times which are 1-definable from y and such that

$$(x, T_1, \ldots, T_n, \ldots) \equiv_1 (y, S_1 \ldots, S_n, \ldots).$$

This fact, together with Theorem 5B.4 for $\equiv_1$, lets us conclude that y is a local martingale.   ⊣

At this point we can mention several research projects that would be interesting to carry out.

(1) Identify which constructions carried out in hyperfinite adapted spaces are of a "definable character." This idea has a purpose behind it. Suppose you prove a result in a hyperfinite space, where things are easier in principle. If you know that the procedure you followed is definable, you can conclude that the result holds in all adapted spaces. A good example: Itô's formula can be proved in a fairly easy way in a hyperfinite adapted space (see Keisler [1988] or Albeverio, Fenstad, Hoegh-Krohn, and Lindstrøm [1986]), and one can then conclude that it is true in all adapted spaces. *Does this mean that the construction in the easy hyperfinite proof was in some sense definable?* We can ask the same question for many other results. *Can we use easy definable hyperfinite constructions to prove theorems such as the Doob-Meyer decomposition or the Girsanov formula?*

(2) One question suggested by the proof of Proposition 5B.7: *Which concepts in the general theory of processes are definable or have definable versions?* A concept

such as local martingale postulates the existence of a sequence of localizing times. Durrett's result quoted in the proof shows that there is a definable localizing sequence. How pervasive is this phenomenon?

CHAPTER 6

ELEMENTARY EXTENSIONS

Up to this point we have been working with stochastic processes that live on fixed adapted spaces. The only time we considered an extra factor was in the preceding chapter when we studied intrinsic filtrations. In this chapter we are going to allow changes in the adapted space in which we are working. We will see how one can handle these changes within the theory of adapted distributions, and look at some of the constructions that are carried out in the general theory of processes through the eyes of a model theorist.

Suppose we are given an adapted space $\Omega = (\Omega, \mathcal{F}_t, P)$. How do probabilists modify it? We list some of the ways this is done, and give intuitive explanations of why they could be interesting. The book Jacod [1979] contains a good presentation of these topics.

Filtrations were introduced into probability theory in order to capture the information available at each time t, so it is natural to consider a situation where different observers have different ways of gathering that information. For example, an observer who has more powerful tools than another should be able to collect more information. This means, of course, that he has a bigger filtration! Similarly, if two observers have equipment which is not compatible and look at different aspects of the world, they should end up with information (i.e., filtrations) that are not necessarily comparable.

Formally, these comments lead us to consider the following ways to change a given adapted space $(\Omega, \mathcal{F}_t, P)$ to a new adapted space $(\Omega, \mathcal{G}_t, Q)$. In each case, the sample set Ω stays the same.

I. Restricting the filtration: $\mathcal{G}_t \subseteq \mathcal{F}_t$ for each t, and $Q \subseteq P$.

II. Extending the filtration: $\mathcal{G}_t \supseteq \mathcal{F}_t$ for each t, and $Q \supseteq P$.

Another possible change in an adapted space is to replace the probability measure P by another probability measure Q on the same set of events. This can be viewed as a different way of assigning probabilities to events. A special case of this situation is where $Q \ll P$ (that is, Q is absolutely continuous with respect to P). Formally, this means:

III. Changing the probability measure: $\mathcal{G}_t = \mathcal{F}_t$ for each t, but $Q \neq P$.

Another way of changing an adapted space is suggested by work in stochastic differential equations, particularly with regard to the different types of solutions that are studied in the literature (see Jacod [1979]). They lead us to consider enriching an adapted spaces in order to accommodate more measurable sets, so that there will be more stochastic processes on the new space. In changes of this kind, the sample set Ω may be replaced. One common way of doing this is by considering products of adapted spaces as follows.

IV. Given $(\Omega, \mathcal{F}_t, P)$ and $(\Gamma, \mathcal{G}_t, Q)$. define the product as the new adapted space $(\Omega \times \Gamma, \mathcal{H}_t, R)$ where $\mathcal{H}_t$ is the smallest filtration satisfying the usual conditions and such that for each t, $\mathcal{H}_t \supseteq \mathcal{F}_t \times \mathcal{G}_t$. and R is a probability measure on $\Omega \times \Gamma$ which has P and Q as marginals.

There is a natural way to fit each of the original adapted spaces Ω and Γ within the product space Λ, so that Ω and Γ can be gotten from Λ by restricting the filtration. We will do this using the concept of a filtration isomorphism introduced in Definition 5A.4. There is a filtration isomorphism between Ω and the restriction of Λ to the first coordinate events, and between Γ and the restriction of Λ to the second coordinate events.

In this chapter we examine some of these new structures from the model theoretic point of view. This will lead us to the concepts of elementary extension. elementary embedding, and elementary chain. In the last section of this chapter we will use an elementary chain to construct a saturated adapted space. This construction will show that every adapted space has a saturated elementary extension.

6A. Restricting the filtration

Let us first study the case where we restrict the original filtration $(\mathcal{F}_t)$. In general the question of taking an arbitrary subfiltration is so broad than it is almost impossible to say anything interesting. For this reason, we will concentrate on special cases that have been studied in the literature (see Chapter 9 in Jacod [1979]).

Consider the case where we have an adapted process x on our original adapted space $(\Omega, \mathcal{F}_t, P)$. Suppose that we now restrict the filtration $(\mathcal{F}_t)$ to a filtration $\mathcal{H}_t$ such that x remains adapted with respect to $\mathcal{H}_t$. This is the least one could ask for, since the minimum amount of information that one must have in a filtration should be what can be observed from the process x at any given time s (i.e.. the x_t's with $t \leq s$). If we now look at the process x as an object in the new adapted space $\Omega' = (\Omega, \mathcal{H}_t, P)$, we can ask what properties of x are preserved in moving from the first adapted to the second.

Do we always have $(\Omega, x) \equiv (\Omega', x)$? The answer is "No." In Jacod [1979], Exercise 9.1. there is an example that shows that the local martingale property and other properties are not preserved under this kind of change in the filtration. Since we know that the local martingale property is preserved under $\equiv$ (even under $\equiv_1$). then the answer has to be "No." In the other hand there are some important properties which are preserved. such as being a martingale, a quasi-martingale, or a semimartingale (see Jacod [1979]). As an illustration, here is a proof that the martingale property is preserved under this change in the adapted space.

PROPOSITION 6A.1. *Suppose x is a martingale with respect to $(\Omega, \mathcal{F}_t, P)$. $\mathcal{H}_t \subseteq \mathcal{F}_t$ for each t, and x is adapted with respect to $(\Omega, \mathcal{H}_t, P)$. Then x is a martingale with respect to $(\Omega, \mathcal{H}_t, P)$.*

PROOF. The integrability conditions have nothing to do with the filtration. So we just have to check that for all s and t, if $s \leq t$ then $E[x_t|\mathcal{H}_s] = x_s$ a.s. The following string of equalities is clear.

$$E[x_t|\mathcal{H}_s] = E[E[x_t|\mathcal{F}_s]|\mathcal{H}_s] = E[x_s|\mathcal{H}_s] = x_s. \qquad \dashv$$

This proposition and the preceding results lead us to propose a problem with a theme that we have not yet mentioned. Probabilists in general care about a few properties which are important in their subject, but there are many other properties that they may not have considered, or which they would regard as irrelevant. On the other hand, from the point of view of a model theorist it is natural to consider all properties of a structure.

PROBLEM 6A.2. *Identify those properties of a stochastic process x on an adapted space $(\Omega, \mathcal{F}_t, P)$ that are preserved under restricting to an adapted space $(\Omega, \mathcal{H}_t, P)$ such that $\mathcal{H}_t \subseteq \mathcal{F}_t$ for each t and x is adapted with respect to $(\Omega, \mathcal{H}_t, P)$.*

This problem is similar in spirit to the preservation theorems of first order model theory, where for example it is proved that a theory is preserved under unions of chains if and only if it has an $\forall\exists$ set of axioms Chang and Keisler [1990]). Proposition 6A.1 and Jacod [1979] should give us an idea of what to expect.

6B. Extending the filtration

We now turn our attention to another situation that can be treated quite nicely within our framework. In the paper Fajardo [1990a] there is an informal discussion of how a model theoretic approach can naturally take us to a probabilistic concept. This seems to be an appropriate place to recall the argument in Fajardo [1990a], since it gives us the opportunity to see an example of how model theory and stochastic processes can interact.

Model theory has its roots in modern algebra. Many concepts and results from the early years in model theory were clearly motivated by problems from algebra (Chang and Keisler [1990]). Sometimes, when building the model theory of a subject that has very little to do with algebra, one is led to notions that are quite unnatural in the field under study. At the same time, many ideas from basic model theory have turned out to have a wider range of applicability. One such concept is the notion of an elementary extension. What could an elementary extension of an adapted space be?

We proceed informally (for a more detailed discussion see Fajardo [1990a]). The first thing we have to do is to make precise what we mean by an extension

of a model and what the elements of the universe are. In fact, universes in this
setting are more complicated than in first order logic. The analog of a universe
is an adapted space $\Omega = (\Omega, \mathcal{F}_t, P)$, and an adapted model is a structure (Ω, x)
where x is a stochastic process on Ω. In probability theory the properties of
the elements $\omega \in \Omega$ are irrelevant. The objects that are really important are the
measurable sets in the filtration. These should play the role of elements of the
universe. We can expand the vocabulary for a given adapted space Ω by adding
names for each measurable set. Here this just means that we add its characteristic
function. So we are led to the structure $(\Omega, \mathbb{I}_A)_{A \in \mathcal{F}_1}$. For the next step we
should define what we mean by an extension of an adapted space Ω. The idea
is fairly obvious: an adapted space whose filtration contains more sets at each
time. This can be naturally translated into a structure of the form $\Omega' = (\Omega, \mathcal{G}_t, P)$
with $\mathcal{F}_t \subseteq \mathcal{G}_t$. Finally, we must decide when an adapted model (Ω', x') is an
elementary extension of an adapted model (Ω, x). By analogy with first order
logic, we must decide when the old elements have the same properties with respect
to both adapted models. Here is the formal definition.

DEFINITION 6B.1.   *Let $\Omega = (\Omega, \mathcal{F}_t, P)$ and $\Gamma = (\Gamma, \mathcal{G}_t, Q)$ be adapted spaces.
We write $\Omega \sqsubseteq \Gamma$ if the sample sets Ω and Γ are the same, $\mathcal{F}_t \subseteq \mathcal{G}_t$ for each $t \in [0, 1]$,
and $P[A] = Q[A]$ for each set $A \in \mathcal{F}_1$.*

*Γ is an **elementary extension** of Ω, in symbols $\Omega \prec \Gamma$, if $\Omega \sqsubseteq \Gamma$ and for every tuple
$\mathbb{I}_{A_1}, \ldots, \mathbb{I}_{A_n}$ of characteristic functions of $\mathcal{F}_1$-measurable sets and adapted function
f, we have*

$$f^\Omega(\mathbb{I}_{A_1}, \ldots, \mathbb{I}_{A_n}) = f^\Gamma(\mathbb{I}_{A_1}, \ldots, \mathbb{I}_{A_n}) \ a.s.$$

*An adapted model (Γ, x) is an **elementary extension** of (Ω, x), in symbols
$(\Omega, x) \prec (\Gamma, x)$, if $\Omega \sqsubseteq \Gamma$, x is a stochastic process on Ω, and*

$$f^\Omega(x, \mathbb{I}_{A_1}, \ldots, \mathbb{I}_{A_n}) = f^\Gamma(x, \mathbb{I}_{A_1}, \ldots, \mathbb{I}_{A_n}) \ a.s.$$

for each $\mathbb{I}_{A_1}, \ldots, \mathbb{I}_{A_n}$ and each adapted function f.

*Also, Ω is called an **elementary subspace** of Γ if Γ is an elementary extension of
Ω.*

Here are three easy exercises for the reader.

EXERCISE 6B.2.   *(a) $(\Omega, x) \prec (\Gamma, x)$ implies $(\Omega, x) \equiv (\Gamma, x)$.*
(b) If $\Omega \sqsubseteq \Gamma$ and $\mathcal{G}_t = \mathcal{F}_t$ for all $t < 1$, then $\Omega \prec \Gamma$.
*(c) If $\Omega \prec \Gamma$ then $\mathcal{F}_1 \cap \mathcal{G}_t = \mathcal{F}_t$ for all t. Thus if Γ is a proper elementary
extension of Ω then $\mathcal{G}_1 \neq \mathcal{F}_1$, so Γ has more measurable sets than Ω.*

The following theorem from Fajardo [1990a] gives a characterization of elementary extensions which will be one of our main tools. It basically tells us that
the process x is irrelevant in the definition. All that really matters is the set of
measurable sets and the way they are related at different times. Recall that $P[A|\mathcal{F}]$
denotes the conditional expectation $E[\mathbb{I}_A|\mathcal{F}]$.

THEOREM 6B.3.   *Suppose Ω and Γ are adapted spaces with $\Omega \sqsubseteq \Gamma$. The following
are equivalent:*
(i) $\Omega \prec \Gamma$.

(*ii*) $(\Omega, x) \prec (\Gamma, x)$ *for every stochastic process x on Ω.*

(*iii*) *For every $t \in [0, 1]$ and every $A \in \mathcal{F}_1$, $P[A|\mathcal{F}_t] = P[A|\mathcal{G}_t]$ a.s.*

Proof. It is easily seen that (ii) implies (i) and that (i) implies (iii). We prove that (iii) implies (ii). We have to prove that for every fixed sequence of measurable sets $A_1, \ldots, A_n$ we have

$$(12) \qquad f^{\Omega}(x, \mathbb{I}_{A_1}, \ldots, \mathbb{I}_{A_n}) = f^{\Gamma}(x, \mathbb{I}_{A_1}, \ldots, \mathbb{I}_{A_n}) \ a.s.$$

This is done, as usual, by induction, and the key step is the conditional expectation step. The following argument should already be familiar. Suppose that f is of the form $E[g|t]$. Then by induction hypothesis we have

$$g^{\Omega}(x, \mathbb{I}_{A_1}, \ldots, \mathbb{I}_{A_n}) = g^{\Gamma}(x, \mathbb{I}_{A_1}, \ldots, \mathbb{I}_{A_n}) \ a.s.$$

Observe that $g^{\Omega}(x, \mathbb{I}_{A_1}, \ldots, \mathbb{I}_{A_n})$ is $\mathcal{F}_1$-measurable, and therefore can be approximated as the pointwise limit of a sequence of $\mathcal{F}_1$-simple functions S_n. so that

$$\lim S_n = g^{\Omega}(x, \mathbb{I}_{A_1}, \ldots, \mathbb{I}_{A_n}) \ a.s.$$

Each function S_n can be written as a linear combination of $\sum_0^r c_i \mathbb{I}_{C_i}$ of $\mathcal{F}_1$-measurable sets C_i. Then by (iii) applied to the C_i's, for every t we have

$$E[S_n|\mathcal{F}_t] = E[S_n|\mathcal{G}_t] \ a.s.$$

If we take limits on both sides of this equation we get the desired equation (12)). ⊣

Somewhat surprisingly, and interestingly, condition (iii) in Theorem 6B.3 has an equivalent formulation in purely probabilistic terms (see Jacod [1979]), and there is a name attached to the property. It says that for every t, $\mathcal{F}_1$ and $\mathcal{G}_t$ are conditionally independent given $\mathcal{F}_t$. Here is the formal definition; for more about this notion see Loève [1977-1978] or Dellacherie and Meyer [1978].

Definition 6B.4. *Let $\mathcal{F}$, $\mathcal{G}$ and $\mathcal{H}$ be σ-algebras on a probability space $(\Omega, \mathcal{E}, P)$, i.e., be sub-σ-algebras of $\mathcal{E}$. We say that $\mathcal{F}$ and $\mathcal{G}$ are* **conditionally independent** *given $\mathcal{H}$ if for all $A \in \mathcal{F}$ and $B \in \mathcal{G}$,*

$$P[A \cap B|\mathcal{H}] = P[A|\mathcal{H}] \cdot P[B|\mathcal{H}].$$

Remark 6B.5. *Conditional independence given the trivial σ-algebra $\mathcal{H} = \{\Omega, \emptyset\}$ is just the traditional notion of a pair of independent σ-algebras: For all $A \in \mathcal{F}$ and $B \in \mathcal{G}$, $P[A \cap B] = P[A] \cdot P[B]$.*

The particular case which we are interested in is the following.

Exercise 6B.6. *If $\mathcal{H} \subseteq \mathcal{G}$, then $\mathcal{F}$ and $\mathcal{G}$ are conditionally independent given $\mathcal{H}$ if and only if for all $A \in \mathcal{F}$, $P[A|\mathcal{H}] = P[A|\mathcal{G}]$.*

So, we can write:

Corollary 6B.7. $\Omega \prec \Gamma$ *if and only if $\Omega \sqsubseteq \Gamma$ and for every t, $\mathcal{F}_1$ and $\mathcal{G}_t$ are conditionally independent given $\mathcal{F}_t$.* ⊣

Here are some other characterizations of $\Omega \prec \Gamma$ which follow from Theorem 6B.3.

COROLLARY 6B.8. *Suppose $\Omega \sqsubseteq \Gamma$. The following are equivalent.*
(i) $\Omega \prec \Gamma$.
(ii) For every Ω-adapted process x. $(\Omega. x) \equiv (\Gamma, x)$.
(iii) Every Ω-martingale is a Γ-martingale.

PROOF. By Theorem 6B.3, (i) implies (ii). Proposition 1C.5 shows that (ii) implies (iii). To show that (iii) implies (i), for each $\mathcal{F}_1$-set A consider the Ω-martingale x defined by $x_t = P[A|\mathcal{F}_t]$, $t \in [0, 1]$. Then $x_1 = \mathbb{I}_A$. By (iii), x_t is a Γ-martingale, so

$$P[A|\mathcal{G}_t] = E[\mathbb{I}_A|\mathcal{G}_t] = E[x_1|\mathcal{G}_t] = x_t = P[A|\mathcal{F}_t].$$

This proves condition (iii) of Theorem 6B.3.   ⊣

Can you think of an example where condition (iii) of the preceding corollary holds? Here is a very familiar one.

PROPOSITION 6B.9. *Given an adapted process x on an adapted space Γ, let Γ^x be the adapted space $(\Gamma, \mathcal{I}_t^x, P)$ where $\mathcal{I}_t^x$ is the intrinsic filtration of x with respect to Γ. Then $\Gamma^x \prec \Gamma$. In fact, Γ^x is the $\sqsubseteq$-least $\Omega \prec \Gamma$ such that x is adapted with respect to Ω.*

PROOF. First we show $\Gamma^x \prec \Gamma$. Let y be a martingale on Γ^x. Then y is adapted on Γ^x. so $\Gamma^{x,y} = \Gamma^x$. By Proposition 5A.3 (a), $(\Gamma^{x,y}, x, y) \equiv (\Gamma, x, y)$. Consequently. $(\Gamma^x, x, y) \equiv (\Gamma, x, y)$ and $(\Gamma^x, y) \equiv (\Gamma, y)$. It follows that $\Gamma^x \prec \Gamma$.

Now, let $\Omega = (\Gamma, \mathcal{F}_t, P)$ be an adapted space $\Omega \prec \Gamma$ such that x is adapted on Ω. By Theorem 6B.3, we have $(\Omega, x) \prec (\Gamma. x)$. By definition, this means that for all $f \in AF$, $f^\Omega(x) = f^\Gamma(x)$ a.s. In particular.

$$E[g^\Omega(x)|\mathcal{F}_t] = E[g^\Gamma(x)|\mathcal{G}_t] \, a.s.$$

for every adapted function g. By Proposition 5A.3 (c) it follows that $\mathcal{I}_t^x \subseteq \mathcal{F}_t$, so $\Gamma^x \sqsubseteq \Omega$.   ⊣

It is important to point out that Hoover [1992] studied the relation $\Omega \prec \Gamma$ from another point of view, as we will see in the next section.

Inspired by the model theory of first order logic, the next natural step is to talk about elementary embeddings. We will make use of the notion of a filtration isomorphism introduced in Definition 5A.4. An important thing to keep in mind about filtration isomorphisms is that they ignore individual points in the sample space, and are mappings between events modulo null sets.

DEFINITION 6B.10. *Let Ω and Γ be adapted spaces. We say that h is an elementary embedding from Ω into Γ, in symbols $h : \Omega \prec \Gamma$. if h is a filtration isomorphism from Ω to some $\Gamma' \prec \Gamma$.*

COROLLARY 6B.11. *If $h : \Omega \prec \Gamma$ and x is a stochastic process on Ω, then $\tilde{h}(x)$ is defined on Γ and $(\Omega, x) \equiv (\Gamma, \tilde{h}(x))$.*

PROOF. By Proposition 5A.5 and Theorem 6B.3.   ⊣

The following theorem from Fajardo [1990a] gives us the adapted version of a well known result from first order logic.

THEOREM 6B.12. *Let Γ be an adapted space. If Ω is a hyperfinite adapted space in a κ-saturated nonstandard universe for a sufficiently large cardinal κ, then Γ is elementarily embeddable into Ω.*

PROOF. Let $(G_\alpha)_{\alpha \in \kappa}$ be a listing of the σ-algebra $\mathcal{G}_1$ of Γ, and consider the expanded model $(\Gamma, (\mathbb{I}_{G_\alpha})_{\alpha \in \kappa})$. Let Ω be a hyperfinite adapted space in a nonstandard universe which is κ-saturated. Using the strong form of the Adapted Universality Theorem (Theorem 2C.3), find $(z_\alpha)_{\alpha \in \kappa}$ in Ω such that

$$(13) \qquad (\Gamma, (\mathbb{I}_{G_\alpha})_{\alpha \in \kappa}) \equiv (\Omega(z_\alpha)_{\alpha \in \kappa}).$$

Observe that the functions z_α are $\{0, 1\}$-valued functions (i.e., characteristic functions) almost surely. By changing z_α on a set of measure zero we can assume that it is indeed a characteristic function. Let F_α denote the set defined by z_α, so that $z_\alpha = \mathbb{I}_{F_\alpha}$ and we may rewrite (13)) as

$$(14) \qquad (\Gamma, (\mathbb{I}_{G_\alpha})_{\alpha \in \kappa}) \equiv (\Omega, (\mathbb{I}_{F_\alpha})_{\alpha \in \kappa}).$$

Now for $t \in [0, 1]$, let $\mathcal{H}_t$ be the complete σ-algebra generated by $\{F_\alpha : G_\alpha \in \mathcal{G}_t\}$. Using a monotone class argument and the completeness of $\mathcal{F}_t$, it can be proved that $\mathcal{H}_t \subseteq \mathcal{F}_t$ for each t. Again using a monotone class argument, it follows that the function $h(G_\alpha) = F_\alpha$ is a filtration isomorphism from $(\Gamma, \mathcal{G}_t, Q)$ to $(\Omega, \mathcal{H}_t, P)$. We claim that

$$(15) \qquad \forall A \in \mathcal{H}_1, \quad P[A|\mathcal{F}_t] = P[A|\mathcal{H}_t].$$

Once (15) is proved, we can conclude by Theorem 6B.3 that $(\Omega, \mathcal{H}_t, P) \prec (\Omega, \mathcal{F}_t, P)$, and thus h is the desired elementary embedding from Γ into Ω.

To prove (4), first use a monotone class argument to reduce (15)) to the special case that $A = F_\beta$ for some $\beta < \kappa$. Let $u = Q[G_\beta|\mathcal{G}_t]$, which is a random variable in Γ. By Proposition 5A.5, $\tilde{h}(u) = P[F_\beta|\mathcal{H}_t]$. On the other hand, it follows from (14)) that $v = P[F_\beta|\mathcal{F}_t]$. $\dashv$

REMARK 6B.13. *In the preceding theorem, it suffices to take Ω in a nonstandard universe which is λ-saturated where $\lambda \geq \omega_1$ and the σ-algebra $\mathcal{G}_1$ for Γ is generated by a family of cardinality at most λ.*

After seeing the results in this section, a model theorist would immediately ask if there is an analog of the elementary chain theorem of first order logic in the context of stochastic processes and adapted spaces. The answer is "Yes, there is one." We are going to use a slightly more general notion of extension, and for this reason we will wait a little longer before giving this analog.

6C. Amalgamation and adjunction

Now it is time to look at Hoover's work on extending filtrations. The paper Hoover [1992] presented another way of extending an adapted process. He was motivated by the various types of solutions of stochastic differential equations that had been studied in the literature. Let us quote from the introduction of his paper:

"In probability theory, one often argues so:

Enlarging the space, if necessary, there exists a random variable Y such that...

Constructing the extensions is close to trivial, so no more is generally said. In the general theory of processes, in which a filtration is present, some care must be exercised, because recklessly enlarging the filtration may destroy essential properties of the original processes relative to the filtration, such as the martingale property or the Markov property."

In the literature on stochastic differential equations, solutions of the most general kind are called *weak solutions*. Intuitively, weak solutions were introduced because there was not enough room for a solution in the original space where the equation was studied, but by properly enriching the original space, a solution could be found in a bigger space. It is thus natural to see how the nature of the equation is affected by enlarging the original space. Hoover has isolated the key ingredients of this situation and has explained it from the point of view of the adapted distributions. His treatment is phrased in very general terms, but here we will consider the simplest case relevant for our purposes.

An exposition of Hoover's results, including detailed proofs, appears in Emilio Remolina's masters' thesis [R] at the Universidad de Los Andes, written under the direction of the first author of this book. We draw freely from this reference.

Using the fact that hyperfinite adapted spaces are saturated, we can find an explanation for what follows. In general, if we are given processes x and x' such that $(\Omega, x) \equiv (\Gamma, x')$, and then another process y' on Γ, there is no reason to have a process y on Ω such that $(\Omega, x, y) \equiv (\Gamma, x', y')$. But it is natural to ask whether one can enlarge Ω in order to accommodate a new process y so that the desired property holds. Problems of this type are well known in model theory, and it is no surprise that something similar can be done here. We first need some definitions.

DEFINITION 6C.1. *Given adapted spaces $\Omega = (\Omega, \mathcal{F}_t, P)$ and $\Gamma = (\Gamma, \mathcal{G}_t, Q)$, an* **adapted product** *is an adapted space $\Lambda = (\Omega \times \Gamma, \mathcal{H}_t, R)$ where $(\mathcal{H}_t)$ is the smallest filtration satisfying the usual conditions and containing the product σ-algebra $\mathcal{F}_t \times \mathcal{G}_t$ for each t, and R is a probability measure whose marginals are P and Q.*

There is a natural way to identify the space Ω inside Λ. Simply identify the σ-algebras $\mathcal{F}_t$ for Ω with the σ-algebras

$$\mathcal{F}_t \times \Gamma = \{F \times \Gamma : F \in \mathcal{F}_t\}.$$

Similarly, any stochastic process x on Ω has a natural extension to Λ by means of the formula $x(\omega, \gamma, t) = x(\omega, t)$. The next lemma shows that this natural identification is an elementary embedding. The proof is easy and is left as an exercise for the reader.

LEMMA 6C.2. *Let Ω and Γ be adapted spaces and let Λ be an adapted product of Ω and Γ. Then the natural embeddings $j(F) = F \times \Gamma$ and $k(G) = \Omega \times G$ are elementary embeddings $j : \Omega \prec \Lambda$ and $k : \Gamma \prec \Lambda$.* ⊣

The next theorem is analogous to the amalgamation principle for elementary embeddings which is prominent in classical model theory. It is the reason that the adapted product construction is useful.

THEOREM 6C.3. (*Amalgamation Theorem*) *Let adapted spaces* $\Upsilon = (\Upsilon. \mathcal{E}_t, S)$, $\Omega = (\Omega, \mathcal{F}_t, P)$, *and* $\Gamma = (\Gamma, \mathcal{G}_t, Q)$ *be given, with elementary embeddings* $h : \Upsilon \prec \Omega$ *and* $i : \Upsilon \prec \Gamma$. *Then there is an adapted product space* $\Lambda = (\Omega \times \Gamma, \mathcal{H}_t, R)$ *of* $(\Omega, \mathcal{F}_t, P)$ *and* $(\Gamma, \mathcal{G}_t, Q)$, *with natural embeddings* $j : \Omega \prec \Lambda$ *and* $k : \Gamma \prec \Lambda$ *such that (ignoring null sets):*

(i) *For each set* $C \in \mathcal{E}_1$, $j(h(C)) = k(i(C))$.

(ii) *For every process* x *on* Υ, $\widetilde{j}(\widetilde{h}(x)) = \widetilde{k}(\widetilde{i}(x))$.

Note that condition (i) in this theorem says that the elementary embeddings $h \circ j : \Upsilon \prec \Lambda$ and $i \circ k : \Upsilon \prec \Lambda$ are the same, that is, the following diagram is commutative:

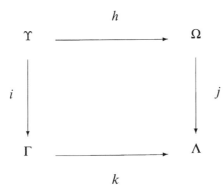

PROOF. With hindsight the probability measure R comes up naturally. We first define a function R_0 on the set of rectangles $A \times B$ where $A \in \mathcal{F}$ and $B \in \mathcal{G}$ by

$$(16) \qquad R_0[A \times B] = \int_\Upsilon \widetilde{h}^{-1}(P[A|h(\mathcal{E}_1)]) \cdot \widetilde{i}^{-1}(Q[B|i(\mathcal{E}_1)]) dS.$$

The product measure theorem (see [As], Theorem 2.6.2) shows that R_0 can be extended to a probability measure R on $\mathcal{F}_1 \times \mathcal{G}_1$.

We now prove that R has property (i). Let $C \in \mathcal{E}_1$ and let D be the complement of C. Then

$$j(h(C)) = h(C) \times \Gamma = (h(C) \times i(C)) \cup (h(C) \times i(D)),$$
$$k(i(C)) = \Omega \times i(C) = (h(C) \times i(C)) \cup (h(D) \times i(C)).$$

So all we have to do is to show that

$$(17) \qquad R[h(C) \times i(D)] = 0 = R[h(D) \times i(C)].$$

Using (16), we have

$$\widetilde{h}^{-1}(P[h(C)|h(\mathcal{E}_1)]) = \mathbb{I}_C, \widetilde{i}^{-1}(P[i(D)|i(\mathcal{E}_1)]) = \mathbb{I}_D,$$

$$R[h(C) \times i(D)] = \int_\Upsilon \mathbb{I}_C \cdot \mathbb{I}_D \, dS = \int_\Upsilon \mathbb{I}_{C \cap D} \, dS = 0.$$

This proves the first equation in (17), and the proof of the second equation is similar. Part (ii) follows easily from (i).   ⊣

The following exercise gives an another property of the above adapted product construction which is interesting in its own right but will not be needed later on.

EXERCISE 6C.4.   *In the preceding theorem, the σ-algebras $\mathcal{F}_1 \times \Gamma$ and $\Omega \times \mathcal{G}_1$ with the measure R are conditionally independent given $h(\mathcal{E}_1) \times \Gamma$ and $\Omega \times i(\mathcal{E}_1)$.*

In an adapted product $\Lambda = \Omega \times \Gamma$, we would like to simplify things by identifying Ω and Γ with their images $j(\Omega)$ and $k(\Gamma)$, so that Ω and Γ are elementary subspaces of Λ. But the Amalgamation Theorem becomes very confusing if one tries to state it with this simplification (part (i) seems to say $C = C$, and part (ii) seems to say $x = x$). Instead, we will adopt a compromise that works out well. We'll identify the first coordinate Ω with its image $j(\Omega)$, so that $\Omega \prec \Lambda$, but will still distinguish the second coordinate Γ from its image $k(\Gamma)$. We can do this easily by broadening our notion of an elementary extension.

DEFINITION 6C.5.   *Given two adapted spaces Ω and Λ, we say that Λ is an **elementary extension** of Ω if $\Lambda = \Omega \times \Gamma$ for some nonempty set Γ, and $j : \Omega \prec \Lambda$ where j is the mapping $j(C) = C \times \Gamma$.*

We will identify the set Ω with a Cartesian product $\Omega \times \{\gamma\}$ of Ω with a one-element set. Thus any elementary extension of Ω in the original sense of Definition 6B.1 is also an elementary extension in the new sense of Definition 6C.5 with a one-element set Γ. We will use the notation $\Omega \prec \Lambda$ and the phrase "elementary subspace" as one would expect. If $\Omega \prec \Lambda$ in this sense, we can safely identify each process x on Ω with its image $\widetilde{j}(x)$ on Λ.

The following theorem is a consequence of the Amalgamation Theorem and the Intrinsic Isomorphism Theorem 5A.10. It is a perfect companion to the Definability Theorem 5B.3 in the same chapter.

THEOREM 6C.6.   *(Adjunction Theorem) Let x and x' be processes on adapted spaces Ω and Γ such that $(\Omega, x) \equiv (\Gamma, x')$, and let y' be another process on Γ. Then there is an elementary extension Λ of Ω and a stochastic process y on Λ such that $(\Lambda, x, y) \equiv (\Gamma, x', y')$.*

PROOF.   Let $(\mathcal{I}_t)$ and $(\mathcal{J}_t)$ be the intrinsic filtrations of x and x'. By Proposition 6B.9 we know that $(\Omega, \mathcal{I}_t, P) \prec (\Omega, \mathcal{F}_t, P)$ and $(\Gamma, \mathcal{J}_t, Q) \prec (\Gamma, \mathcal{G}_t, Q)$. The Intrinsic Isomorphism Theorem 5A.10 gives us a filtration isomorphism i from $(\Omega, \mathcal{I}_t, P)$ to $(\Gamma, \mathcal{J}_t, Q)$ such that $\widetilde{i}(x) = x'$. Let $\Upsilon = (\Omega, \mathcal{I}_t, P)$. Then $\Upsilon \prec \Omega$ and $i : \Upsilon \prec \Gamma$. The Amalgamation Theorem now gives us an elementary extension Λ of Ω with an elementary embedding $k : \Gamma \prec \Lambda$ such that $\widetilde{k}(x') = \widetilde{k}(\widetilde{i}(x)) = x$. Let $y = \widetilde{k}(y')$, which is a process on Λ. Then

$$(\Gamma, x', y') \equiv (\Lambda, \widetilde{k}(x'), \widetilde{k}(y')) \equiv (\Lambda, x, y). \qquad \dashv$$

With the Adjunction Theorem and the notions introduced in this chapter at hand we can include two results, also due to Hoover [1992], which show the usefulness of the notion of an elementary extension. The first result is another characterization of the adapted distribution, and the second is the elementary chain theorem in a very general form. We need a new definition.

DEFINITION 6C.7.   *Given two adapted models* (Ω, x) *and* (Γ, x'). *The relation* $\hookrightarrow$ *is defined as follows:*

$(\Omega, x) \hookrightarrow (\Gamma, x')$ *if and only if for every cadlag* Ω-*adapted stochastic process* y, *there exists an elementary extension* Λ *of* Γ *and a* Λ-*adapted process* y' *such that* $(\Omega, x, y) \equiv_0 (\Lambda, x', y')$.

Notice the similarity of this notion with the finite embeddability relation $(\Omega, x) \rightarrow_0 (\Gamma, x')$ from Chapter 4. The next theorem gives a characterization of adapted equivalence in terms of the relation $\hookrightarrow$.

THEOREM 6C.8. $(\Omega, x) \equiv (\Gamma, x')$ *if and only if* $(\Omega, x) \hookrightarrow (\Gamma, x')$ *and* $(\Gamma, x') \hookrightarrow (\Omega, x)$.

PROOF. The implication from left to right is a corollary of the Adjunction Theorem 6C.6. The proof of the other direction is just like the proof of the analogous result for finite embeddability, Theorem 6B.3, and is left as an exercise for the reader.   ⊣

The final result in this chapter is the adapted analog of a well known theorem in model theory, the Elementary Chain Theorem (see Chang and Keisler [1990], Theorem 3.19). In this context Hoover calls the result the Extension Chain Theorem. We give a proof with details taken from Remolina [1993].

To get the notion of an elementary chain, we will iterate the notion of an elementary extension. Recall that an elementary extension of an adapted space Ω is an adapted space whose sample set is a Cartesian product $\Omega \times \Gamma$ for some nonempty set Γ. When we iterate this we'll get adapted spaces whose sample sets are Cartesian products of the form $\Omega \times \prod_{\alpha \leq \beta} \Gamma_\alpha$, identifying Ω with $\Omega \times \Gamma_0$ where Γ_0 is a space with a single point.

DEFINITION 6C.9.   *Let* κ *be a limit ordinal, and let* $\Omega^\alpha = \left(\Omega^\alpha, \mathcal{F}_t^\alpha, P^\alpha \right)$, $\alpha < \kappa$, *be an increasing chain of adapted spaces. The chain is said to be an* **elementary chain** *if for each* $\alpha \leq \beta < \kappa$, $\Omega^\alpha \prec \Omega^\beta$, *and* Ω^β *has a sample set of the form* $\Omega^\beta = \prod_{\alpha \leq \beta} \Gamma^\alpha$.

The **union** *of the elementary chain is the adapted space* $\Lambda = (\lambda, \mathcal{H}_t, R) = \bigcup_{\alpha < \kappa} \Omega^\alpha$ *where* Λ *is the set* $\prod_{\alpha < \kappa} \Gamma^\alpha$, $\mathcal{H}_t$ *is the smallest filtration containing each* $\mathcal{F}_t^\alpha$ *and satisfying the usual conditions, and* R *is the unique probability measure on* $\mathcal{H}_1$ *which agrees with each* P^α *on* $\mathcal{F}_1^\alpha$.

THEOREM 6C.10.   (*Elementary Chain Theorem*) *The union of an elementary chain of adapted spaces is an elementary extension of each adapted space in the chain.*

PROOF. Let Ω^α, $\alpha < \kappa$ be an elementary chain with union Λ. Let $\alpha < \kappa$ and $B \in \mathcal{F}_1^\alpha$. We want to prove that given $t \in [0, 1]$, $P[B | \mathcal{F}_t^\alpha] = P[B | \mathcal{H}_t]$. It will then follow from Theorem 6B.3 that $\Omega^\alpha \prec \Lambda$.

Let $x = P[B|\mathcal{H}_t]$. We wish to show that $x = P[B|\mathcal{F}_t^\alpha]$. By Theorem 6B.3 we have

$$P[B|\mathcal{F}_t^\alpha] = P[B|\mathcal{F}_t^\beta] = E[x|\mathcal{F}_t^\beta],$$

whenever $\alpha < \beta < \kappa$. Suppose first that κ has cofinality ω, that is, there is an increasing sequence β_n, $n \in \mathbb{N}$ of ordinals which converges to κ. Then $E[x|\mathcal{F}_t^{\beta_n}]$ converges to x almost surely (see, for example, Theorem 35.5 of Ethier and Kurtz [1986]). It follows that $x = P[B|\mathcal{F}_t^\alpha]$ as required.

Now suppose that κ has uncountable cofinality. The random variable x is a limit of a countable sequence of simple functions x_n. For each n, there is a $\beta < \kappa$ such that x_n is $\mathcal{F}_t^\beta$-measurable. By uncountable cofinality there is a $\gamma < \kappa$ such that each x_n is $\mathcal{F}_t^\gamma$-measurable. Therefore x is $\mathcal{F}_t^\gamma$-measurable, so again

$$x = E[x|\mathcal{F}_t^\gamma] = P[B|\mathcal{F}_t^\alpha]. \qquad \dashv$$

One way of building saturated models in classical model theory is by means of elementary chains. As could be expected, here we can build saturated adapted spaces (Hoover [1992]) using elementary chains. We will do this in the next section.

6D. Saturation from elementary chains

One of the main contributions of the model theoretic view of stochastic processes that we have developed in this book is the appearance of saturated adapted spaces. Our hyperfinite adapted spaces constructed in Chapter 2 are the prime examples of saturated adapted spaces. We established some of their properties in Chapter 3, and have seen how these properties are keys to the nonstandard approach to stochastic analysis.

The construction of saturated models in first order logic is well known. Two ways of building saturated models are via ultraproducts and via elementary chains (see Chang and Keisler [1990]). Since our approach to nonstandard analysis is the superstructure one, which is closely related to the ultraproduct construction, we can say that the hyperfinite models correspond to the ultraproduct method of building saturated models. This section contains a construction due to Hoover [1992] that is the analog of the elementary chain method.

We now proceed to sketch the construction in Hoover [1992] of a saturated adapted space via elementary chains. For fully detailed proofs of his results see López [1989] and Remolina [1993] (we follow this reference). The proof is divided into three natural stages. Our goal is the following theorem.

THEOREM 6D.1. *Every adapted space has a saturated elementary extension.*

We begin with a definition.

DEFINITION 6D.2. *Let x be a stochastic process on an adapted space $\Omega = (\Omega, \mathcal{F}_t, P)$. Ω is said to be x-saturated if for any processes x' and y' on another space $(\Gamma, \mathcal{G}_t, Q)$ such that $x \equiv x'$, there is a process y on Ω such that $(x, y) \equiv (x', y')$.*

Observe that Ω is saturated if and only if Ω is x-saturated for every stochastic process x on Ω.

LEMMA 6D.3. *Let Ω be an adapted probability space. Then for each stochastic process x on Ω, there is an x-saturated elementary extension of Ω.*

PROOF. The set AF of adapted functions has the cardinality $\mathfrak{c}$ of the continuum. Therefore the set of all $\equiv$-classes of pairs of processes (x', y') with $x' \equiv x$ can be listed in a sequence of length $2^{\mathfrak{c}}$. That is, there is a sequence of $2^{\mathfrak{c}}$ adapted models $((\Gamma_\alpha, x_\alpha, y_\alpha) : \alpha < 2^{\mathfrak{c}})$ such that $x_\alpha \equiv x$ for each α, and for every (x', y') with $x' \equiv x$ there is an $\alpha < 2^{\mathfrak{c}}$ such that $(x_\alpha, y_\alpha) \equiv (x', y')$. Construct an elementary chain $(\Omega_\alpha : \alpha \leq 2^{\mathfrak{c}})$ of adapted spaces such that $\Omega_0 = \Omega$, and for each $\alpha < 2^{\mathfrak{c}}$ there is a process z_α on $\Omega_{\alpha+1}$ with $(x, z_\alpha) \equiv (x_\alpha, y_\alpha)$.

To construct Ω_β we use the Adjunction Theorem when β is a successor ordinal $\beta = \alpha + 1$, and the Elementary Chain Theorem when β is a limit ordinal. The union of the chain, $\Omega^{2^{\mathfrak{c}}}$, is the desired x-saturated elementary extension of Ω. ⊣

The next lemma is trivial, so we omit the proof.

LEMMA 6D.4. *Let Ω be an adapted space. If x is a stochastic process on Ω, and Ω is x-saturated, then any elementary extension of Ω is x-saturated.* ⊣

LEMMA 6D.5. *Every adapted space Ω has an elementary extension Γ such that Γ is x-saturated for every process x on Ω.*

PROOF. Let $(x^\alpha : \alpha < 2^{\mathfrak{c}})$ be a list of all stochastic processes on Ω, up to $\equiv$-equivalence. Using Lemma 6D.3 at successor stages and the Elementary Chain Theorem at limit stages, construct an elementary chain $(\Omega^\alpha : \alpha \leq 2^{\mathfrak{c}})$ of adapted spaces such that $\Omega^0 = \Omega$, and $\Omega^{\alpha+1}$ is x^α-saturated for each $\alpha < 2^{\mathfrak{c}}$. Then by the Elementary Chain Theorem and the preceding lemma, $\Gamma = \Omega^{2^{\mathfrak{c}}}$ has the required property. ⊣

The last step in the construction is Theorem 6D.1.

PROOF. of Theorem 6D.1: Using the preceding lemmas, construct an elementary chain $(\Lambda_\alpha : \alpha \leq \mathfrak{c}^+)$ of adapted spaces such that $\Lambda_0 = \Omega$ and for each $\alpha < \mathfrak{c}^+$, $\Lambda_{\alpha+1}$ is x-saturated for every process x on Λ_α. The union of the chain, $\Lambda_{\mathfrak{c}^+}$, is an elementary extension of Ω by the Elementary Chain Theorem. We claim that $\Lambda_{\mathfrak{c}^+}$ is saturated. To see this, let x be a stochastic process on $\Lambda_{\mathfrak{c}^+}$. Then for each $r \in \mathbb{R}$ and $t \in [0, 1]$, the set $\{\omega : x_t(\omega) \leq r\}$ is measurable in Λ_α for some $\alpha = \alpha(r, t) < \mathfrak{c}^+$. Since the successor cardinal $\mathfrak{c}^+$ is regular, the set of ordinals $\{\alpha(r, t) : r \in \mathbb{R}, t \in [0, 1]\}$ has an upper bound $\beta < \mathfrak{c}^+$. It follows that x is a process on Λ_β. Therefore $\Lambda_{\mathfrak{c}^+}$ is x-saturated, and the claim is proved. ⊣

We now have a construction of a saturated adapted space that does not use methods from nonstandard analysis. The hyperfinite adapted space is not only saturated but has the stronger property of being universal and homogeneous. This leads us to the following problem.

PROBLEM 6D.6. *Find a construction of a universal homogeneous adapted space which does not use methods from nonstandard analysis.*

In the last chapter of this book. Chapter 9, we will encounter another case where extending the filtration results in a saturated adapted space. There, instead of starting with an arbitrary adapted space we will start with an adapted space that is already saturated. Moreover, we will only add sets to the filtration which belong to $\mathcal{F}_1$, that is, are measurable in the original space. We saw early in this chapter, in Exercise 6B.2, that an adapted space formed in this way will never be an elementary extension of the original space. Instead, the objective will be to show that the new space is still saturated.

But first we need to add to our toolkit one more method of building saturated adapted spaces. We have seen in Chapter 2 that hyperfinite adapted spaces are saturated. In the next chapter we will show that any Loeb adapted space which satisfies the adapted analog of being atomless is saturated.

CHAPTER 7

RICH ADAPTED SPACES

In the preceding chapters we have often used the adjective "rich" to describe the powerful properties of saturated and hyperfinite adapted spaces. In this chapter we formally introduce the mathematical notion of a rich adapted space and prove three main theorems about them (in these theorems we allow filtrations which are not necessarily right continuous): (1) With a countable time line, an adapted space is rich if and only if it is saturated. (2) Every atomless adapted Loeb space is rich. (3) For every rich adapted space with the real time line, the corresponding right continuous adapted space is saturated. This will show that all atomless adapted Loeb spaces are saturated. In the next chapter we will see that rich adapted spaces have some strong properties which quickly lead to a variety of applications in probability theory.

Many applications of nonstandard analysis take the following form, sometimes called the lifting procedure. Start with a problem stated in standard terms. Lift everything in sight up to the nonstandard world. Construct a sequence of internal approximate solutions of some kind indexed by the natural numbers. Use the Saturation Principle to extend the sequence through the hyperintegers. Finally, take object number H where H is a sufficiently small infinite hyperinteger, and come back down to the standard world to get a solution of the original problem. Nonstandard arguments of this sort are similar to standard compactness arguments, but can work in cases where a compactness argument fails. There are several examples of the lifting procedure in Chapter 2.

The lifting procedure is well known to workers in the field. Unfortunately, to those mathematicians not familiar with the theory, the beauty and power of these techniques remains a complete mystery. One way to address this issue is to try to convince experts of the value of nonstandard analysis, so that they are encouraged to study it and then be able to use it. One tries to present interesting results so that the experts in the field will dare to look closely at them. But predictably, these efforts have encountered resistance. For a related discussion of these issues consult Keisler [1991], Keisler [1994].

In a rich adapted space the lifting procedure has a short cut which is defined in purely standard terms. This short cut is a countable compactness principle for basic sections, which are sets of random variables obtained from adapted functions. It makes it possible to enjoy many of the benefits of nonstandard methods without paying the normal entry fee.

Rich adapted spaces originally arose out of a much more general theory developed in the papers Keisler [1991], Fajardo and Keisler [1996a]–Fajardo and Keisler [1995], Keisler [1995]–Keisler [1997a]. We will concentrate here on those aspects which are most closely related to the model theoretic concepts we have introduced concerning adapted distributions. The treatment given here is the result of several stages of evolution which will be easier to describe at the start of the next chapter.

We will do some groundwork in Section 7A, where we state some results we will need about the law function, notably the Skorokhod Representation Theorem. In Section 7B we make further preparations, defining the notion of an atomless adapted space and extending the results of Section 7A to them. Rich adapted spaces are defined in Section 7C. The three theorems mentioned in the first paragraph of this Chapter will be proved in Sections 7C, 7D, and 7E respectively.

7A. The law function

We begin with some notation for spaces of random variables and measures associated with a given probability space Ω. We will then collect some facts that explain the behavior of the law function when defined on random variables on an atomless probability space. We already did something in this direction in Chapter 1, but here we will go further.

Given two metric spaces $(\mathcal{M}, \rho)$ and $(\mathcal{N}, \sigma)$, the product metric is the metric space $(\mathcal{M} \times \mathcal{N}, \rho \times \sigma)$ where

$$(\rho \times \sigma)((x_1 x_2), (y_1, y_2)) = \max(\rho(x_1, y_1), \sigma(x_2, y_2)).$$

We will occasionally need to form countable products of metric spaces $\mathcal{M} = \Pi_n \mathcal{M}_m$. If $\mathcal{M}_n$ has the metric ρ_n for each n, the product metric will be taken to be

$$\rho(x, y) = \Sigma_n \min(1, \rho_n(x_n, y_n))/2^n.$$

This sum always converges and gives $\mathcal{M}$ the product topology.

The distance between a point $x \in \mathcal{M}$ and a nonempty set $C \subseteq \mathcal{M}$ is

$$\rho(x, C) = \inf\{\rho(x, y) : y \in C\}.$$

Let $M = (M, \rho)$ and $N = (N, \sigma)$ be Polish spaces. Given a probability space $\Omega = (\Omega, \mathcal{G}, P)$, let $\mathcal{M} = L^0(\Omega, M)$ be the set of all random variables on Ω with values in M, (that is, P-measurable functions from Ω into M, identifying functions which are equal P-almost surely). ρ_0 is the metric of convergence in probability on $\mathcal{M}$.

$$\rho_0(x, y) = \inf\{\varepsilon : P[\rho(x(\omega), y(\omega)) \leq \varepsilon] \geq 1 - \varepsilon\}.$$

Our convention will be that, given a probability or adapted space Ω and a Polish space M or N, we use the corresponding script letter for the space of random variables $\mathcal{M} = L^0(\Omega, M)$ or $\mathcal{N} = L^0(\Omega, N)$. We identify each $m \in M$ with the random variable in $\mathcal{M}$ with the constant value m. With this identification, M is a closed subspace of $\mathcal{M}$. The space of all real valued random variables on Ω will be denoted by $\mathcal{R}$.

The space of Borel probability measures on M with the Prohorov metric

$$d(\mu, \nu) = \inf\{\varepsilon : \mu(K) \leq \nu(K^\varepsilon) + \varepsilon \text{ for all closed } K \subseteq M\}$$

is denoted by $Meas(M)$. It is again a Polish space, and convergence in $Meas(M)$ is the same as weak convergence. Each measurable function $x : \Omega \to M$ induces a measure $law(x) \in Meas(M)$, and the function

$$law : \mathcal{M} \to Meas(M)$$

is continuous. Moreover, if the measure P is atomless, then for each M the function law maps the set of all $x \in \mathcal{M}$ onto $Meas(M)$. We refer to the book Ethier and Kurtz [1986] for background on these matters.

PROPOSITION 7A.1. *Let Ω be an atomless probability space. Then*
(a) For each continuous function $h : M \to N$ there is a continuous function

$$\bar{h} : law(\mathcal{M}) \to law(\mathcal{N})$$

such that

$$\bar{h}(law(x)) = law(\widehat{h}(x)),$$

where $\widehat{h} : \mathcal{M} \to \mathcal{N}$ is defined by $(\widehat{h}(x))(\omega) = h(x(\omega))$, i.e., the following diagram is commutative:

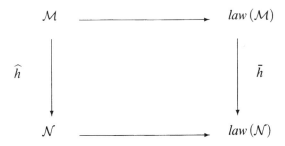

(b) law maps $\mathcal{M}$ continuously onto $Meas(M)$. In particular, the image $law(\mathcal{M})$ is closed.

(c) (Skorokhod Representation Theorem). For every random variable x on Ω with values in a Polish space M and sequence (c_n) which converges to $law(x)$ in $Meas(M)$, there exists a sequence (x_n) in $\mathcal{M}$ such that $x_n \to x$ and $law(x_n) = c_n$.

(d) Let $x, y \in \mathcal{M}, z \in \mathcal{N}$, and $law(x) = law(y)$. Then there is a sequence $z_n \in \mathcal{N}$ such that $law(x, z_n)$ converges to $law(y, z)$.

PROOF. We leave (a) and (b) as exercises. Part (c) is a well-known and frequently used result in probability theory (e.g., see Ethier and Kurtz [1986]). We prove (d). Let (x_n) be a sequence of simple functions converging to x in $\mathcal{M}$. Then $law(x_n) \to law(x)$. By part (c) there is a sequence (y_n) converging to y in $\mathcal{M}$ such that $law(x_n) = law(y_n)$ for each n. Then each y_n is simple. By Proposition 3B.3 (the weak form of saturation for atomless probability spaces), there is a sequence (z_n) in $\mathcal{N}$ such that $law(x_n, z_n) = law(y_n, z)$ for each n. We have $(y_n, z) \to (y, z)$ in $\mathcal{M} \times \mathcal{N}$, and therefore $law(x_n, z_n) \to law(y, z)$. Moreover, since $x_n \to x$ in $\mathcal{M}$, $d(law(x_n, z_n), law(x, z_n)) \to 0$. Therefore $law(x, z_n) \to law(y, z)$. $\dashv$

7B. Atomless adapted spaces

In this section we introduce the important notion of an atomless $\mathbb{L}$-adapted space, and discuss some of its properties.

From this point on in the book, we shall often use adapted spaces indexed by time sets which are different than the usual $[0, 1]$, and will also drop the requirement that the filtration be right continuous. Let $(\mathbb{L}, \leq)$ be a linearly ordered set with a first element 0 and a last element 1. We shall usually take $\mathbb{L}$ to be either a finite set $\mathbb{F} \subseteq [0, 1]$ or the set $\mathbb{B}$ of dyadic rationals in $[0, 1]$. We say that

$$\Omega = (\Omega, (\mathcal{G}_t)_{t \in \mathbb{L}}, P)$$

is an $\mathbb{L}$-**adapted space** if P is a complete probability measure on $\mathcal{G}_1$, $\mathcal{G}_t$ is a P-complete σ-algebra for each $t \in \mathbb{L}$, and $\mathcal{G}_s \subseteq \mathcal{G}_t$ whenever $s < t$ in $\mathbb{L}$.

An $\mathbb{L}$-adapted space Ω is **universal** if for every random variable x on some other $\mathbb{L}$-adapted space there is a random variable y on Ω such that $x \equiv_\mathbb{L} y$. Note that we said "random variable" in this definition, not "stochastic process". The back and forth property and the notion of being saturated are defined in the analogous way for $\mathbb{L}$-adapted spaces.

We emphasize that we are making a distinction between an $\mathbb{L}$-adapted space and a just plain adapted space. An $\mathbb{L}$-adapted space has the time index set $\mathbb{L}$ and its filtration $(\mathcal{G}_t)$ is not necessarily right continuous. On the other hand, an adapted space always has the real time index set $[0, 1]$ and its filtration $(\mathcal{F}_t)$ is always right continuous.

Note that every adapted spaces is a $[0, 1]$-adapted space. An adapted space Ω has the back and forth property as an adapted space if and only if it has the back and forth property as a $[0, 1]$-adapted space. However, the property of Ω being universal as a $[0, 1]$-adapted space is stronger than being universal as an adapted space, and the property of Ω being saturated as a $[0, 1]$-adapted space is stronger than being saturated as an adapted space.

A probability space (without the filtration) is just the special case of an $\mathbb{L}$-adapted space where $\mathbb{L} = \{1\}$ is a singleton (we allow $0 = 1$ in this case). From here on we can define notions for probability spaces and adapted spaces together by giving a definition for $\mathbb{L}$-adapted spaces in general.

The following result is proved in the same way as the Saturation Theorem.

THEOREM 7B.1. *An $\mathbb{L}$-adapted space is saturated if and only if it is universal and has the back and forth property.* $\dashv$

Definition 7B.2. *On an* $\mathbb{L}$*-adapted space* Ω*, a sequence of stochastic processes* x_n *is said to* **converge in** $\mathbb{L}$**-adapted distribution** *to* x*, in symbols* $x_n \Rightarrow_{\mathbb{L}} x$*. if*

$$\lim_{n \to \infty} law\,(x_n) = law\,(x)$$

and

$$\lim_{n \to \infty} E[f(x_n)] = E[f(x)]$$

for every $\mathbb{L}$*-adapted function* f*.*

We now define the notion of an atomless $\mathbb{L}$-adapted space.

Definition 7B.3. *Let* $\mathcal{E}$ *and* $\mathcal{F}$ *be* σ*-algebras with* $\mathcal{E} \subseteq \mathcal{F}$*.* $\mathcal{F}$ *is said to be atomless over* $\mathcal{E}$ *if for every* $U \in \mathcal{F}$ *of positive probability, there is a set* $V \subseteq U$ *in* $\mathcal{F}$ *such that*

$$0 < P[V|\mathcal{E}] < P[U|\mathcal{E}]$$

on a set of positive probability. Following Hoover and Keisler [1984]*, we say that an* $\mathbb{L}$*-adapted space* Ω *is* **atomless** *if* $(\Omega, \mathcal{G}_0, P)$ *is an atomless probability space, and* $\mathcal{G}_t$ *is atomless over* $\mathcal{G}_s$ *whenever* $s < t$ *in* $\mathbb{L}$*.*

An adapted space is atomless if it is atomless as a $[0,1]$*-adapted space.*

Observe that for any atomless $\mathbb{L}$-adapted space Ω, the corresponding $\mathbb{K}$-adapted space is atomless for every $\mathbb{K} \subseteq \mathbb{L}$. Also, an $\mathbb{L}$-adapted space is atomless if and only if for every finite $\mathbb{F} \subseteq \mathbb{L}$, the corresponding $\mathbb{F}$-adapted space is atomless.

Proposition 7B.4. *Any adapted space which has an adapted Brownian motion is atomless.*

Proof. For any $s < t$ in $[0,1]$, the increment $b(t) - b(s)$ of the Brownian is $\mathcal{F}_t$-measurable and is independent of $\mathcal{F}_s$. This gives enough room to ensure that the adapted space is atomless.   ⊣

Many of the commonly used ordinary adapted spaces are atomless as adapted spaces.

Example 7B.5. *Let* M *have more than one element. Each of the commonly used ordinary adapted spaces*

$$C([0,1], M), D([0,1], M), L^0([0,1], M), L^p([0,1], M)$$

listed in Example 1E.9 *is atomless as an adapted space (for some Borel measure).*

We now compare the properties of being atomless and universal.

Theorem 7B.6. *Let* $\mathbb{L} \subseteq \mathbb{R}$*. Every universal* $\mathbb{L}$*-adapted space is atomless.*

Proof. Let b be a Brownian motion on an adapted space Γ, and let $\Gamma_{\mathbb{L}}$ be the $\mathbb{L}$-adapted space formed by restricting the filtration of Γ. Suppose Ω is a universal $\mathbb{L}$-adapted space. Then there is a process w on Ω with $w \equiv_{\mathbb{L}} b$. Whenever $s < t$ in $\mathbb{L}$, the increment $w(t) - w(s)$ is $\mathcal{F}_t$-measurable, independent of $\mathcal{F}_s$, and has positive expected value. It follows that Ω is an atomless $\mathbb{L}$-adapted space.   ⊣

We know From Theorem 1E.2 that a plain probability space is universal if and only if it is atomless. Here is the corresponding result for adapted spaces with finite time sets.

Theorem 7B.7. *Let $\mathbb{F}$ be a finite linear ordering. Any atomless $\mathbb{F}$-adapted space is universal.*

Proof. The proof is in Keisler [1997a], Proposition 6.8. The details are intricate, but the basic idea is as follows. First, introduce a stronger $\mathbb{F}$-adapted analog of the notion of a simple random variable. Among other things, the $\mathbb{F}$-simple random variables are dense in each $\mathbb{M}$. The main step is to show that for every random variable x' on some $\mathbb{F}$-adapted space, the given atomless $\mathbb{F}$-adapted space Ω has a sequence of $\mathbb{F}$-simple random variables x_n such that $x_n \Rightarrow_{\mathbb{F}} x'$ and x_n is a Cauchy sequence in $\mathcal{M}$. It then follows that the $x = \lim_{n \to \infty} x_n$ exists in $\mathcal{M}$ and $x \equiv_{\mathbb{F}} x$.   ⊣

Proposition 7A.1 also carries over to adapted spaces with finite time sets. The following results are taken from the paper Keisler [1997a], Propositions 6.7, 7.3, and 7.4. Parts (a)–(d) of this theorem are the analogs of parts (a)–(d) of Proposition 7A.1 respectively. We omit the proofs, which again use the adapted analog of a simple random variable.

Proposition 7B.8. *Let $\mathbb{F}$ be finite and let Ω be an atomless $\mathbb{F}$-adapted space.*

(a) For each continuous function $h : M \to N$. if $x_n \Rightarrow_{\mathbb{F}} x$ in $\mathcal{M}$ then $\widehat{h}(x_n) \Rightarrow_{\mathbb{F}} \widehat{h}(x)$ in $\mathcal{N}$.

(b) (Closure property) Any sequence x_n of random variables on Ω such that $\{law(x_n) : n \in \mathbb{N}\}$ is contained in a compact set has a subsequence y_n such that $y_n \to_{\mathbb{F}} y$ for some random variable y on Ω.

(c) (Skorokhod property) If x, y_n are random variables on Ω such that $y_n \Rightarrow_{\mathbb{F}} x$, there exist random variables x_n on Ω such that $x_n \equiv_{\mathbb{F}} y_n$ for each n and x_n converges to x in probability.

(d) (Density property) If x, y, z are random variables on Ω such that $x \equiv_{\mathbb{F}} y$. then there is a sequence of random variables z_n on Ω such that $(x, z_n) \Rightarrow_{\mathbb{F}} (y, z)$. ⊣

Our next result shows that it is not so easy to find universal $\mathbb{L}$-adapted spaces when $\mathbb{L}$ is a dense set of times.

Theorem 7B.9. *No ordinary $\mathbb{B}$-adapted space is universal.*

Proof. We modify the proof of Theorem 1E.11. Suppose Ω is a universal $\mathbb{B}$-adapted space. Then for each $u \in (0,1]$ there must be a set A^u of measure $1/2$ which belongs to $\mathcal{G}_t$ for all $t \geq u$ and is independent of $\mathcal{G}_s$ for all $s < u$. Then $\{A^u : u \in (0,1]\}$ is an uncountable independent family of sets. Thus by Lemma 1E.10. Ω cannot be ordinary.   ⊣

7C. Richness versus saturation

In this section we introduce the notion of a rich $\mathbb{L}$-adapted space and prove the key theorem that when $\mathbb{L}$ is countable, an $\mathbb{L}$-adapted space is rich if and only if it is countable. We will see later that this characterization which is very convenient for applications.

A family F of sets is said to be **countably compact** if for any countable set $G \subseteq F$, if every finite $H \subseteq G$ has a nonempty intersection then G has a nonempty intersection.

Definition 7C.1. *Let*

$$\Omega = (\Omega, (\mathcal{G}_t)_{t \in \mathbb{L}}, P)$$

be an $\mathbb{L}$-adapted space, and let M be Polish. A set B will be called a **subbasic set** *(in $\mathcal{M}$ over $(\Omega, \mathbb{L})$) if*

$$B = \{x \in \mathcal{M} : law(x) \in C, E[f(x)] \in D\}$$

for some $\mathbb{L}$-adapted function f and compact sets $C \subseteq Meas(M)$ and $D \subseteq \mathbb{R}$.
    A **basic set** *(in $\mathcal{M}$ over $(\Omega, \mathbb{L})$) is a finite union of subbasic sets.*
    A set A is called a **basic section** *(in $\mathcal{M}$ over $(\Omega, \mathbb{L})$) if $A = \{x \in \mathcal{M} : (x, z) \in B\}$ for some basic set B in $\mathcal{M} \times \mathcal{N}$ over $(\Omega, \mathbb{L})$ and some $z \in \mathcal{N}$.*
    An $\mathbb{L}$-adapted space Ω is **rich** *if it is atomless and for each $\mathcal{M}$ the family of basic sections in $\mathcal{M}$ over $(\Omega, \mathbb{L})$ is countably compact.*

We remark that if Ω is a rich $\mathbb{K}$-adapted space, then whenever $\{0, 1\} \subseteq \mathbb{K} \subseteq \mathbb{L}$, the restriction of Ω to $\mathbb{K}$ is a rich $\mathbb{K}$-adapted space.

The above definition of a rich $\mathbb{L}$-adapted space is simpler than the definition in Fajardo and Keisler [1996b], but the paper Keisler [1997a] shows that the two definitions are equivalent. We will prove that rich $\mathbb{L}$-adapted spaces exist in the next section. As a warmup, we prove two easy propositions.

Proposition 7C.2. *For each Polish space M, there is a countable set of basic sets whose union is dense in $\mathcal{M}$.*

Proof. Let $D = \{d_0, d_1, \ldots\}$ be a countable subset of M, and let $D_n = \{d_0, \ldots, d_n\}$. For each n, the set $C_n = law^{-1}(D_n)$ is basic. The set of simple functions with values in D is dense in $\mathcal{M}$, and each simple function with values in D belongs the set C_n for some n. Thus $\bigcup_n C_n$ is dense in $\mathcal{M}$.   ⊣

Proposition 7C.3. *For very basic section A there is a compact set C such that $law(x) \in C$ for all $x \in A$ (so A is contained in the subbasic set $\{x : law(x) \in C\}$).*

Proof. By definition, each basic section is contained in a set of the form $B = \{x \in \mathcal{M} : law(x, z) \in D\}$ for some compact set $D \subseteq Meas(M \times N)$. By Proposition 7A.1 (a), the projection function $f : M \times N \to M$ induces a continuous function $g : Meas(M \times N) \to Meas(M)$. Thus the set $g(D)$ is compact in $Meas(M)$ and $A \subseteq \{x \in (M) : law(x) \in g(D)\}$.   ⊣

We now come to the main theorem of this section.

Theorem 7C.4. *Suppose $\mathbb{L}$ is countable and there exists a rich $\mathbb{L}$-adapted space. Then an $\mathbb{L}$-adapted space Ω is saturated if and only if it is rich.*

Proof. By Corollary 3D.3, for each Polish space N we may choose a countable $\equiv_\mathbb{L}$-dense set $\{h_k : k \in \mathbb{N}\}$ of $\mathbb{L}$-adapted functions on N-valued random variables.
    Suppose first that Ω is saturated. Then Ω is atomless by Theorems 7B.1 and 7B.6. Let

$$A_k = \{x \in \mathcal{M} : (x, z_k) \in B_k\}$$

be a countable sequence of basic sections in $\mathcal{M}$, where each B_k is a basic set in $\mathcal{M} \times \mathcal{N}_k$ and $z_k \in \mathcal{N}_k$. Assume that any finite number of A_k's is nonempty. By induction on k, one can replace B_k by one of its subbasic components and still retain the property that any finite number of the A_k's is nonempty. (Alternatively, use the Koenig tree theorem). We may therefore assume without loss of generality that each B_k is a subbasic set,

$$B_k = \{ y \in \mathcal{M} \times \mathcal{N}_k : law\,(y) \in C_k, E[f_k(y)] \in D_k \}$$

where C_k, D_k are compact and f_k is an $\mathbb{L}$-adapted function. Let N be the product of the Polish spaces N_k and let $z = \langle z_0, z_1, \ldots \rangle \in \mathcal{N}$. Then there are compact sets E_k and $\mathbb{L}$-adapted functions g_k such that for each k,

$$A_k = \{ x \in \mathcal{M} : law\,(x, z) \in E_k, E[g_k(x, z)] \in D_k \}.$$

For each k, let $x_k \in \bigcap_{m \le k} A_k$, and let $d_k = E[h_k(z)]$, where h_k is from the countable $\equiv_{\mathbb{L}}$-dense set. Then whenever $m \le k$ we have

$$law\,(x_k, z) \in E_m, E[g_m(x_k, z)] \in D_m, E[h_m(z)] = d_m.$$

By hypothesis there is a rich $\mathbb{L}$-adapted space Γ. Γ is atomless, so by Proposition 7B.8 (a), Γ is $\mathbb{F}$-universal for each finite $\mathbb{F} \subseteq \mathbb{L}$. Then for each k there exists (x'_k, z'_k) on Γ such that for all $m \le k$,

$$law\,(x'_k, z'_k) \in E_m, E[g_m(x'_k, z'_k)] \in D_m, E[h_m(z'_k)] = d_m.$$

Thus the intersection of any finite number of the basic sets

$$F_m = \{ (x', z') : law\,(x', z') \in E_m, E[g_m(x', z')] \in D_m, E[h_m(z')] = d_m$$

is nonempty. Since Γ is rich, there is a pair of random variables (x', z') on Γ such that $(x', z') \in \bigcap_m F_m$. Because h_k is $\equiv_{\mathbb{L}}$-dense, we have $z' \equiv_{\mathbb{K}} z$. Then since Ω is saturated there is a random variable x on Ω such that $(x, z) \equiv_{\mathbb{L}} (x', z')$. Therefore $x \in \bigcap_k A_k$, and hence Ω is rich.

Now assume that Ω is rich. We first show that Ω is universal. Let x' be an N-valued random variable on some $\mathbb{L}$-adapted space. For each k let B_k be the basic set

$$B_k = \{ x \in \mathcal{N} : law\,(x) = law\,(x'), E[h_k(x)] = E[h_k(x')] \}.$$

Ω is atomless, so by Proposition 7B.8 (a), Ω is $\mathbb{F}$-universal for each finite $\mathbb{F} \subseteq \mathbb{L}$. Therefore each finite intersection $\bigcap_{m \le k} B_m$ is nonempty. Since Ω is rich, there exists $x \in \bigcap_k B_k$. Then $x \equiv_{\mathbb{L}} x'$, and hence Ω is universal.

By Theorem 7B.1, it now suffices to prove that Ω has the back and forth property. Let x, y, z be random variables on Ω with $x \equiv_{\mathbb{L}} y$ and $z \in \mathcal{M}$. Let $\mathbb{F}$ be a finite subset of $\mathbb{L}$. Since Ω is atomless, Theorem 7B.6 (d) gives us a sequence z_n of random variables on Ω such that $(x, z_n) \Rightarrow_{\mathbb{F}} (y, z)$. Since $\mathbb{L}$ is countable, it follows that there is a sequence z_n such that $(x, z_n) \Rightarrow_{\mathbb{L}} (y, z)$. Let C be the compact set

$$C = \{ law\,(x, z_n) : n \in \mathbb{N} \}.$$

For each $k, m \in \mathbb{N}$, let A_{km} be the basic section

$$A_{km} = \{u \in \mathcal{M} : law\,(x, u) \in C, d\,(law\,(x, u), law\,(y, z)) \leq 1/m,$$

$$|E[f_k(x, u)] - E[f_k(y, z)]| \leq 1/m\}.$$

Then for each k, m we have $z_n \in A_{km}$ for all sufficiently large n. Since Ω is rich, there exists $u \in \bigcap_{m,k} A_{km}$. It follows that $(x, u) \equiv_{\mathbb{L}} (y, z)$, so Ω is saturated.   ⊣

Note that the above proof only needed the existence of an $\mathbb{L}$-adapted space which is rich in the weak sense that the family of basic sets on each $\mathcal{M}$ is countably compact (rather than the family of basic sections). Moreover, in the case that $\mathbb{L}$ is finite, the assumption that there exists a rich $\mathbb{L}$-adapted space can be avoided entirely by using Theorem 7B.8 (b).

7D. Loeb $\mathbb{L}$-adapted spaces are rich

The time has come to prove that there are rich $\mathbb{L}$-adapted spaces for each countable $\mathbb{L}$. This will establish the hypothesis of Theorem 7C.4, so we can conclude that the rich and saturated $\mathbb{L}$-adapted spaces are the same. It should be no surprise that all $\mathbb{L}$-adapted Loeb spaces are rich and saturated.

In this section we let $(\mathbb{L}, \leq)$ be a subset of the hyperreal unit interval $^*[0, 1]$ which contains 0 and 1.

For example, $\mathbb{L}$ can be a finite subset of $^*[0, 1]$, the countable discrete set $\{1 - 1/n : 0 < n \in \mathbb{N}\} \cup \{0, 1\}$, the set $\mathbb{B}$ of dyadic rationals, or the standard unit internal $[0, 1]$.

In Chapter 2 we defined adapted Loeb space. We now need the corresponding $\mathbb{L}$-adapted notion.

DEFINITION 7D.1. *Let*

(18) $$(\Omega, (\bar{\mathcal{G}}_t)_{t \in \{^*[0, 1]\}}, \bar{P})$$

be an internal $^*[0, 1]$-*adapted space. The corresponding* **Loeb $\mathbb{L}$-adapted space** *is the $\mathbb{L}$-adapted space*

(19) $$(\Omega, (\mathcal{G}_t)_{t \in \mathbb{L}}, P)$$

where $P = L(\bar{P})$ *is the Loeb measure of* $\bar{P}$ *and* $\mathcal{G}_t = L(\bar{\mathcal{G}}_t)$ *is the σ-algebra generated by the internal algebra* $\bar{\mathcal{G}}_t \vee \mathcal{N}$ *for each* $t \in \mathbb{L}$.

Thus a Loeb $\mathbb{L}$-adapted space is obtained from a Loeb $^*[0, 1]$-adapted space by restricting the time set to $\mathbb{L}$. Except in trivial cases, the filtration $\mathcal{G}_t$ in a Loeb $\mathbb{L}$-adapted space will not be right continuous.

Throughout this section, it will be understood that Ω is the $\mathbb{L}$-adapted Loeb space 7D.1 (19) corresponding to the $\bar{\mathbb{L}}$-adapted space 7D.1 (18). As in the preceding section, M will always denote a Polish space, and $\mathcal{M} = L^0(\Omega, M)$.

Remember that $\bar{P}$ takes values in the hyperreal unit interval $^*[0, 1]$, while P is a σ-additive probability measure in the usual sense with values in $[0, 1]$. We let $\bar{E}[X]$ be the internal integral of X with respect to the *probability measure $\bar{P}$.

DEFINITION 7D.2.   *A function $X : \Omega \to {}^*M$ will be called $\bar{\mathcal{G}}_t$-measurable if X is internal and $X^{-1}(U) \in \bar{\mathcal{G}}_t$ for each *open set $U \subseteq {}^*M$.*
 Let $\bar{\mathcal{M}}$ be the *metric space of all $\bar{\mathcal{G}}_1$-measurable functions $X : \Omega \to {}^*M$ with the *metric

$$\bar{\rho}_0(X, Y) = {}^* \inf\{\varepsilon : \bar{P}[{}^*\rho(X(\omega), Y(\omega)) \geq \varepsilon] \leq \varepsilon\}.$$

In the literature, $\bar{\mathcal{M}}$ is usually denoted by $SL^0(\Omega, M)$.
 *We say that $X \in \bar{\mathcal{M}}$ is a **lifting** of a function $x : \Omega \to M$, in symbols ${}^oX = x$, if $X(\omega)$ has standard part $x(\omega) \in M$ for P-almost all $\omega \in \Omega$.*
 *Given a set $C \subseteq \mathcal{M}$, the **monad** of C is the set*

$$monad(C) = \{X \in \bar{\mathcal{M}} : {}^oX \in C\}.$$

*The monad of the whole space $\mathcal{M}$ is denoted by $ns^0(\Omega, M)$, and is the set of all X which are liftings of some element of $\mathcal{M}$. The elements $X \in ns^0(\Omega, M)$ are said to be **nearstandard**.*
 *A function $x \in \mathcal{M}$ is **simple** if it has finite range, and a function $X \in \bar{\mathcal{M}}$ is ***simple** if it has *finite range.*

Note that in the space $\bar{\mathcal{M}}$, the distance between any two points is at most one. so all points belong to the same galaxy.

DEFINITION 7D.3.   *Given a set $C \subseteq \mathcal{M}$ and a function $f : C \to \mathcal{N}$, a **uniform lifting** of f is an internal function $F : \bar{\mathcal{M}} \to \bar{\mathcal{N}}$ such that ${}^oF(X) = f({}^oX)$ for all X in the monad of C. If f has a lifting. it is said to be **uniformly liftable**.*

We need the following well known fundamental result in Loeb measure theory.

PROPOSITION 7D.4.   (*López* [1989]]. *Anderson* [1976]) *(a) For any $x : \Omega \to M$, the following are equivalent:*
 (i) *x is Loeb measurable, i.e., $x \in \mathcal{M}$.*
 (ii) *x has a lifting $X \in \bar{\mathcal{M}}$.*
 (iii) *For each infinite $n \in {}^*\mathbb{N}$, x has a *simple lifting $X \in \bar{\mathcal{M}}$ whose range has *cardinality at most n.*
 (iv) *x is a limit of simple functions with respect to ρ_0.*
 (b) Whenever $x, y \in \mathcal{M}$, $x = {}^oX$, and $y = {}^oY$, we have

$$\rho_0(x, y) = {}^o\bar{\rho}_0(X, Y),$$

so that $\bar{\rho}_0$ is a uniform lifting of ρ_0.
 (c) If $X \in \mathcal{R}$ has a finite bound and $x = {}^oX$. then $E[x] = {}^o\bar{E}[X]$.   ⊣

Among other things, this shows that if $X \in \bar{\mathcal{M}}$ is a lifting of some function x, then x is Loeb measurable and thus X is near-standard.

COROLLARY 7D.5.   *(a) Let $t \in \mathbb{L}$. A random variable $x \in \mathcal{M}$ is $\mathcal{G}_t$-measurable if and only if x has a $\bar{\mathcal{G}}_t$-measurable lifting.*
 (b) If $X \in \bar{\mathcal{R}}$ has a finite bound and $x = {}^oX$, then $\bar{E}[X|\bar{\mathcal{G}}_t]$ is a lifting of $E[x|\mathcal{G}_t]$.

PROOF.   For (a), apply Proposition 7D.4 to the $\mathbb{L}$-adapted Loeb space Ω_t formed by redefining $\bar{\mathcal{G}}_s = \bar{\mathcal{G}}_t$ whenever $t \leq s \leq 1$.

For (b), let $y = E[x|\mathcal{G}_t]$, $Y = \bar{E}[X|\bar{\mathcal{G}}_t]$. By (a), y has a $\bar{\mathcal{G}}_t$-measurable lifting Z. For each set $A \in \bar{\mathcal{G}}_t$, we have $A \in \mathcal{G}_t$. By Proposition 7D.4,

$$\bar{E}_A[Y] = \bar{E}_A[X] \approx E_A[x] = E_A[y] \approx \bar{E}_A[Z].$$

Therefore $Y(\omega) \approx Z(\omega) a.s.$, and hence Y lifts y. $\dashv$

We now prove a key lifting theorem for the *law* function.

PROPOSITION 7D.6. (*Keisler* [1991], *Propositions* 2.4 *and* 2.7) *The function law from $\mathcal{M}$ into Meas(M) has a uniform lifting LAW from $\bar{\mathcal{M}}$ into *Meas(M). Moreover, for each $X \in \bar{\mathcal{M}}$, we have $X \in ns^0(\Omega, M)$ if and only if $LAW(X)$ is nearstandard in *Meas(M).*

PROOF. Define the internal function LAW by $LAW(X)(C) = \bar{P}[X^{-1}(C)]$ for each *Borel set C in *M. The family of finite intersections of sets of the form $\{\mu : E_\mu[\psi] \in (a,b)\}$, where $\psi : M \to \mathbb{R}$ is bounded continuous, forms an open basis for *Meas(M). For any such ψ we have

$$E_{law(x)}[\psi] = E[\psi(x)] \approx \bar{E}[(^*\psi)(X)] = \bar{E}_{LAW(X)}[^*\psi(X)].$$

Thus, if $E_{law(x)}[\psi] \in (a,b)$, then $\bar{E}_{LAW(X)}[^*\psi(X)] \in {}^*(a,b)$. It follows that $LAW(X)$ belongs to the star of any open neighborhood of $law(x)$, that is, $LAW(X) \approx law(x)$.

Suppose $LAW(X)$ is near-standard, and let μ be its standard part. For any bounded continuous function $\psi : M \to \mathbb{R}$, we have $\bar{E}_{LAW(X)}[^*\psi] \approx E_\mu[\psi]$. It follows that for each compact set $K \subseteq M$, $\bar{P}[X \in {}^*K^{1/n}] \geq \mu(K) - 1/n$ for all n. Since $^*K \subseteq monad(K)$, $P[X \in monad(K)] \geq \mu(K)$. By Prohorov's Theorem, $\{\mu\}$ is tight, so for each n there is a compact set K_n such that $\mu(K_n) \geq 1 - 1/n$. Therefore $X(\omega)$ is near-standard for P-almost all ω, and $X \in ns^0(\Omega, M)$. $\dashv$

We also need a lifting theorem for adapted functions.

PROPOSITION 7D.7. *Every $\mathbb{L}$-adapted function f from $\mathcal{M}$ to $\mathcal{R}$ has a uniform lifting F from $\bar{\mathcal{M}}$ to $\bar{\mathcal{R}}$.*

PROOF. As in the Adapted Lifting Theorem in Chapter 2, we let F be the internal function which is built in the same way as f but using $^*\varphi$ instead of a continuous function φ and the taking internal conditional expectations $\bar{E}[\cdot|\bar{\mathcal{G}}_t]$ instead of $E[\cdot|\mathcal{G}_t]$. The proof is by induction on complexity. Since each adapted function has a uniform finite bound, we may apply Corollary 7D.5 at the conditional expectation step. $\dashv$

The above result is very much like the Adapted Lifting Theorem from Chapter 2, except that it applies to $\mathbb{L}$-adapted Loeb spaces and adapted functions rather than hyperfinite adapted spaces and conditional processes. In Chapter 2 we said that the Adapted Lifting Theorem is not true for adapted functions, and this was the main reason for introducing conditional processes. Why happened to make the argument work for adapted functions now? The point is that we are now working with an $\mathbb{L}$-adapted space, where $\mathcal{G}_t$ is the σ-algebra generated by an internal algebra $\bar{\mathcal{G}}_t$, while in Chapter 2 we were working with an adapted Loeb space where $\mathcal{F}_t$ was not of this form. The filtration $\mathcal{F}_t$ was right continuous, and $\mathcal{F}_t$ was the σ-algebra generated by $\bigcup_{s \approx t} \bar{\mathcal{G}}_s = \bigcap_{\circ s > t} \bar{\mathcal{G}}_s$.

PROPOSITION 7D.8. *For each basic set B in $\mathcal{M}$, there is a countable decreasing chain of internal sets $B_n \subseteq \bar{\mathcal{M}}$ such that $monad(B) = \bigcap_n B_n$.*

PROOF. It suffices to prove the result for a subbasic set

$$B = \{x \in \mathcal{M} : law(x) \in C, E[f(x)] \in D\}$$

where f is an adapted function and C, D are compact. By the preceding propositions, LAW uniformly lifts law and f has a uniform lifting F. For each n Let B_n be the internal set

$$B_n = \{X \in \bar{\mathcal{M}} : LAW(X) \in {}^*C^{1/n}, \bar{E}[F(X)] \in {}^*D^{1/n}\}.$$

Then $B_1 \supseteq B_2 \supseteq \cdots$. We show that $monad(B) = \bigcap_n B_n$.

Suppose first that $X \in monad(B)$. so that ${}^\circ X = x \in B$. Fix an $n \in \mathbb{N}$. Then $LAW(X) \approx law(x) \in C$, so $LAW(X) \in {}^*C^{1/n}$, and $\bar{E}[F(X)] \approx E[f(x)] \in D$, so $\bar{E}[F(X)] \in {}^*D^{1/n}$. Therefore $monad(B) \subseteq B_n$.

Now suppose $X \in \bigcap_n B_n$. By overspill, there exists $c \in {}^*C$ such that $c \approx LAW(X)$, and $d \in {}^*D$ such that $\bar{E}[F(X)] \approx d$. By a classical result of Robinson, if C is compact then every element of ${}^*C$ is near-standard and its standard part is in C. Therefore $LAW(X)$ is near-standard. By Proposition 7D.6, X is near-standard, and thus has a standard part $x \in \mathcal{M}$. It follows that $LAW(X) \approx law(x) \in C$ and $\bar{E}[F(X)] \approx E[f(x)] \in D$. This shows that $x \in B$ and so $X \in \bigcap_n B_n \subseteq monad(B)$.   ⊣

COROLLARY 7D.9. *For each basic section A in $\mathcal{M}$, there is a countable decreasing chain of internal sets $A_n \subseteq \bar{\mathcal{M}}$ such that $monad(A) = \bigcap_n A_n$.*

PROOF. Let $A = \{x \in \mathcal{M} : (x, z) \in B\}$ where B is basic, and $monad(B) = \bigcap_n B_n$. Take a lifting Z of z. Then $monad(A) = \bigcap_n A_n$ where A_n is the internal set $A_n = \{X \in \bar{\mathcal{M}} : (X, Z) \in B_n\}$.   ⊣

We are now ready to show that every atomless $\mathbb{L}$-adapted Loeb space is rich.

THEOREM 7D.10. *Let $\mathbb{L} \subseteq {}^*[0, 1]$. Every atomless $\mathbb{L}$-adapted Loeb space Ω is rich. Thus rich $[0, 1]$-adapted spaces exist.*

PROOF. Let $C_m, m \in \mathbb{N}$ be a countable decreasing chain of nonempty basic sections in $\mathcal{M}$. By the preceding corollary. for each m there is a countable decreasing chain of internal sets A_{mn} such that $monad(C_m) = \bigcap_n A_{mn}$. By the Saturation Principle, there is an $X \in \bigcap_m(\bigcap_n A_{mn})$, and hence ${}^\circ X \in \bigcap_m C_m$. This shows that the family of basic sections in $\mathcal{M}$ is countably compact. and hence Ω rich.   ⊣

COROLLARY 7D.11. *(a) For each countable $\mathbb{L}$, an $\mathbb{L}$-adapted space is rich if and only if it is saturated.*
(b) Every atomless Loeb probability space is rich.
(c) For each countable $\mathbb{L}$, every atomless $\mathbb{L}$-adapted Loeb space is saturated.   ⊣

Now let's take a look at our hyperfinite adapted spaces, and see when they satisfy the hypothesis of Theorem 7D.10. Since we slightly change the conditions and notation from Chapter 2, we restate the definition of a hyperfinite adapted space.

Let Ω_0 be a hyperfinite set with more than one element, let N be an infinite hyperinteger, and let $\mathbb{B}_N$ be the hyperfinite set of all multiples of 2^{-N} between 0 and 1. Then $\mathbb{B} \subseteq \mathbb{B}_N \subseteq {}^*[0, 1]$. Let $\Omega = \Omega_0^{\mathbb{B}_N}$ be the hyperfinite set of all internal functions from $\mathbb{B}_N$ into Ω_0, let $\bar{\mathcal{G}}$ be the algebra of all internal subsets of Ω, and let $\bar{P}$ be the counting probability measure on $\bar{\mathcal{G}}$, which gives each element of Ω the same weight. If $\omega, \upsilon \in \Omega$ and $t \in {}^*[0, 1]$, we write $\omega \sim_t \upsilon$ if $\omega(s) = \upsilon(s)$ for all $s \leq t$ in $\mathbb{B}_N$. Pick an infinitesimal $\iota \in \mathbb{B}_N$ and let $\bar{\mathcal{G}}_0$ be the algebra of all internal subsets U of Ω such that U is closed under $\sim_\iota$. For $\iota < t \leq 1$ in $^*[0, 1]$ let $\bar{\mathcal{G}}_t$ be the algebra of all internal subsets U of Ω such that U is closed under the equivalence relation $\sim_t$. Since $\bar{\mathcal{G}}$ is the algebra of all internal subsets of Ω, $SL^0(\Omega, M)$ is the set of all internal functions $X : \Omega \to {}^*M$.

DEFINITION 7D.12. *Let* $\mathbb{L} \subseteq {}^*[0, 1]$. *By the* **hyperfinite $\mathbb{L}$-adapted space** *determined by* Ω_0, N, *and* ι *we mean the* $\mathbb{L}$*-adapted Loeb space* $(\Omega, \mathcal{G}_t)_{t \in \mathbb{L}.P}$ *defined above.*

One may wonder why we did not simply take $\iota = 0$, so that $\bar{\mathcal{G}}_0$ is just the algebra of all internal U closed under $\sim_0$. The problem with this choice is that it would make the algebra $\mathcal{G}_0$ be finite when Ω_0 is finite. We wish to allow the possibility that the initial probability space $(\Omega, \mathcal{G}_0, P)$ is atomless even when Ω_0 is finite, as in the next proposition.

PROPOSITION 7D.13. *Let* $\mathbb{L} \subseteq [0, 1]$. *Let* Ω *be the hyperfinite* $\mathbb{L}$-*adapted space determined by* Ω_0, N, *and* ι, *where* $2 \leq |\Omega_0| \in {}^*\mathbb{N}$, $N \in {}^*\mathbb{N}$ *is infinite, and* ι *is infinitesimal.*

(a) $\mathcal{G}_t$ *is atomless over* $\mathcal{G}_s$ *for each* $s < t$ *in* $\mathbb{L}$.

(b) *If either* Ω_0 *is infinite or* $N \cdot \iota$ *is infinite then* $(\Omega, \mathcal{G}_0, P)$ *is atomless, and* Ω *is an atomless* $\mathbb{L}$-*adapted space.* $\dashv$

PROOF. (a) follows from the fact that Ω has the uniform counting measure, and $\sim_\iota$ divides each $\sim_s$-equivalence class into infinitely many pieces. The hypotheses of (b) insure that $\sim_\iota$ has infinitely many equivalence classes all of the same hyperfinite size, so $(\Omega, \mathcal{G}_0, P)$ is atomless. $\dashv$

THEOREM 7D.14. *Suppose* $2 \leq |\Omega_0| \in {}^*\mathbb{N}$, $N \in {}^*\mathbb{N}$ *is infinite, and* ι *is infinitesimal, and either* Ω_0 *is infinite or* $N \cdot \iota$ *is infinite. Then for every* $\mathbb{L} \subseteq [0, 1]$, *the hyperfinite* $\mathbb{L}$-*adapted space* Ω *determined by* Ω_0, N, *and* ι *is rich.*

PROOF. By Proposition 7D.13 and Theorem 7D.10. $\dashv$

7E. Loeb adapted spaces are saturated

We have seen that atomless $[0, 1]$-adapted Loeb spaces are saturated. However, these spaces might not satisfy the usual condition that the filtration is right

continuous. In this section we show that the corresponding right continuous
Loeb spaces are also saturated. We begin with a disturbing example.

EXAMPLE 7E.1. *In a rich $\mathbb{B}$-adapted space*

$$\Omega = (\Omega. \mathcal{G}_t. P).$$

the $(\mathcal{G}_t)$ filtration is never right continuous. That is, $\mathcal{G}_t$ is a proper subset of

$$\bigcap\{\mathcal{G}_s : t < s \in \mathbb{B}\}$$

for all $t < 1$ in $\mathbb{B}$.

PROOF. Let M be a compact metric space with at least two elements. For each
$t \in \mathbb{B}$ let C_t be the basic set

$$C_t = \{x \in \mathcal{M} : E[|x - E[x|\mathcal{G}_t]|] = 0\}$$

$$= \{x \in \mathcal{M} : x \text{ is } \mathcal{G}_t\text{-measurable }\}$$

and let D_t be the basic set

$$D_t = \left\{x \in \mathcal{M} : E[|x - E[x|\mathcal{G}_t]|] = \frac{1}{2}\right\}.$$

Let $1 > t \in \mathbb{B}$ and choose a decreasing sequence $s_n \in \mathbb{B}$ such that $\lim_{n\to\infty} s_n = t$.
The intersection of D_t with each of the sets C_{s_n} is nonempty; let $x = \mathbb{I}_A$ where A is
a $\mathcal{G}_{s_n}$-measurable set of measure $1/2$ which is independent of $\mathcal{G}_t$. The sets C_{s_n} are
decreasing with n because s_n is decreasing. Using the hypothesis that Ω is rich,
there is an x such that $x \in D_t \cap \bigcap_n C_{s_n}$. Then $x = \mathbb{I}_A$ where $A \in \bigcap_{s>t} \mathcal{G}_s \setminus \mathcal{G}_t$. ⊣

The above argument also shows that for any rich $[0. 1]$-adapted space, the
filtration is never right continuous. Of course, this just means that there are no
rich adapted spaces with continuous time which satisfy the usual condition that
the filtration is right continuous. By convention, when we say just plain "adapted
space," we mean a $[0. 1]$-adapted space with a right continuous filtration. Thus
there are no rich adapted spaces. To avoid any confusion, we will usually say
"right continuous adapted space" in this section, even though the words "right
continuous" are redundant.

We saw in Chapter 3 that the hyperfinite right continuous adapted spaces
are saturated. Have we reached the end of the story? No. Every $\mathbb{B}$-adapted
space $(\Omega, (\mathcal{G})_t)_{t\in\mathbb{B}}, P)$, and also every $[0. 1]$-adapted space $(\Omega. (\mathcal{G}_t)_{t\in[0.1]}. P)$, has a
corresponding adapted space $(\Omega. (\mathcal{F}_t)_{t\in[0.1]}. P)$ with the right continuous filtration

$$\mathcal{F}_t = \bigcap_{s>t} \mathcal{G}_s. \mathcal{F}_1 = \mathcal{G}_1.$$

We will see that for every rich $[0. 1]$-adapted space, the corresponding right continuous adapted space is saturated. In particular, this shows that every atomless
right continuous adapted Loeb space is saturated. It follows that saturation as a
$[0. 1]$-adapted space implies saturation as a right continuous adapted space. By
the Saturation Theorem, a right continuous adapted space is saturated if and only

if it is universal and has the back and forth property, and similarly for a $[0, 1]$-adapted space. Here is what is actually going on: A right continuous adapted space has the back and forth property as an adapted space if and only if it has the back and forth property as a $[0, 1]$-adapted space. But it is easier to be universal as an adapted space than as a $[0, 1]$-adapted space, because the right continuous adapted spaces are a subcollection of the $[0, 1]$-adapted spaces.

Now that we are looking at right continuous adapted spaces, we will again work with conditional processes. On an adapted space Ω, we say that a sequence of random variables x_n **converges in adapted distribution** to a random variable x, in symbols $x_n \Rightarrow x$, if $law(x_n)$ converges to $law(x)$, and for each k-fold conditional process f, $E[f(x_n)]$ converges to $E[f(x)]$ in the space $L^0([0, 1]^k, \mathbb{R})$.

We state without proof two results about atomless right continuous adapted spaces from Keisler [1997a], Propositions 9.13 and 9.11. Part (a) says that the space is approximately universal. Part (b) is the analog of Proposition 7B.8 (d) for adapted spaces.

PROPOSITION 7E.2. *Let Ω^{rt} be an atomless right continuous adapted space.*

(a) For each random variable x on some right continuous adapted space, there is a sequence of random variables x_n on Ω^{rt} such that $x_n \Rightarrow x$. ⊣

(b) If x, y, z are random variables on Ω^{rt} such that $x \equiv y$, then there is a sequence of random variables z_n on Ω^{rt} such that $(x, z_n) \Rightarrow (y, z)$.

We now come to our main theorem in this section (see Theorem 9.15 of Keisler [1997a]).

THEOREM 7E.3. *If Ω is a rich $[0, 1]$-adapted space, then the corresponding right continuous adapted space Ω^{rt} is saturated.*

PROOF. We first show that Ω^{rt} is universal. By Lemma 1E.6, it suffices to prove that Ω^{rt} is universal for random variables. Let x' be an M-valued random variable on some right continuous adapted space. By Proposition 7E.2 (a), there is a sequence x_n of random variables on Ω^{rt} which converges in adapted distribution to x'. For each conditional process h and random variable y, let $h(y)$ be the value in Ω and $h^{rt}(y)$ be the value in Ω^{rt}. Using the upcrossing inequality for martingales, one can show that by induction on complexity that for each m-fold conditional process h and each y on Ω, $E[h(y)] = E[h^{rt}(y)]$ almost surely in $[0, 1]^m$.

By Proposition 3D.2 there is a countable dense set $\{h_k : k \in \mathbb{N}\}$ of conditional processes. Let h_k be $m(k)$-fold. There is a set $U \subseteq [0, 1]^{m(k)}$ of measure one on which $E[h_k(x_n)] = E[h_k^{rt}(x_n)]$ for all n. Moreover, $E[h_k^{rt}(x_n)]$ converges to $E[h_k^{rt}(x')]$ in the space $L^0([0, 1]^{m(k)})$. By diagonalizing through subsequences of x_n in the usual way, we may assume that for each k, $E[h_k^{rt}(x_n)]$ converges to $E[h_k^{rt}(x')]$ almost surely in $[0, 1]^{m(k)}$. Hence there is a set $V \subseteq [0, 1]^{\mathbb{N}}$ of product measure one such that for each k, $E[h_k^{rt}(x_n)]$ converges to $E[h_k^{rt}(x')]$, and $E[h_k(x_n)] = E[h_k^{rt}(x_n)]$ for all n, everywhere on V. Therefore $E[h_k(x_n)]$ converges to $E[h_k^{rt}(x')]$ everywhere on V. Take a countable dense set $W \subseteq V$. By saturation, there is a random variable x on Ω such that for all k, $E[h_k(x)] =$

$E[h_k^{rt}(x')]$ everywhere on W. Since $E[h_k^{rt}(x)]$ and $E[h_k^{rt}(x')]$ are right continuous and

$$E[h_k^{rt}(x)](\vec{t}) = \lim_{\vec{s}\downarrow\vec{t}} E[h_k(x)](\vec{s}),$$

it follows that for all k we have

(20) $$E[h_k^{rt}(x)] = E[h_k^{rt}(x')]$$

everywhere. Therefore $x \equiv x'$. This shows that Ω^{rt} is universal. By a similar argument using Proposition 7E.2 (b). it can be shown that Ω^{rt} has the back and forth property. Then by the Saturation Theorem. Ω^{rt} is saturated. ⊣

At the end of Chapter 8 we will prove a stronger result from Keisler [1997a], that for every rich $\mathbb{B}$-adapted space the corresponding right continuous adapted space is saturated. The above theorem. however. is good enough to get the result that we want. which is the following corollary.

COROLLARY 7E.4. *Every atomless Loeb adapted space is saturated.*

PROOF. Let Ω^{rt} be the atomless Loeb adapted space obtained from an internal adapted space

$$(\Omega, (\bar{\mathbb{G}}_t)_{t\in {}^*[0,1]}, \bar{P}),$$

and the corresponding $[0, 1]$-adapted Loeb space

$$(\Omega, (\mathbb{G}_t)_{t\in[0,1]}, P).$$

By Theorem 7E.3. Ω is a rich $[0, 1]$-adapted space. Then by Theorem 7D.10, Ω^{rt} is saturated. ⊣

We conclude this chapter with an open problem.

PROBLEM 7E.5. *If Γ is a saturated adapted space. does there necessarily exist a rich $\mathbb{B}$-adapted space or $[0, 1]$-adapted space Ω such that Γ is the right continuous adapted space Ω^{rt} corresponding to Ω?*

CHAPTER 8

ADAPTED NEOMETRIC SPACES

In this chapter we will establish additional properties of rich adapted spaces which can be applied to stochastic analysis. In Sections 8.1 and 8.2 we give an overview of the theory of neometric spaces, culminating in the Approximation Theorem. Section 8.C gives some typical applications of this theorem.

We briefly survey the evolution of the ideas in this chapter. The paper "From discrete to continuous time", Keisler [1991], introduced a forcing procedure, resembling model theoretic forcing (see Hodges [1985]), which reduced statements about continuous time processes to approximate statements about discrete time processes without going through the full lifting procedure. After some refinements, in the series of papers Fajardo and Keisler [1996a] – Fajardo and Keisler [1995] we worked out a new theory, called the theory of neometric spaces, which has the following objectives: "First, to make the use of nonstandard analysis more accessible to mathematicians, and second, to gain a deeper understanding of why nonstandard analysis leads to new existence theorems. The neometric method is intended to be more than a proof technique—it has the potential to suggest new conjectures and new proofs in a wide variety of settings." (From the introduction in Keisler [1995]).

This theory developed an axiomatic framework built around the notion of a neometric space, which is a metric space with a family of subsets called neocompact sets that are, like the compact sets, closed under operations such as finite union, countable intersection, and projection. In particular, for each adapted probability space there is an associated neometric space of random variables, and the adapted space is rich if the family of neocompact sets is countably compact.

In this book we take advantage of the results in the more recent paper Keisler [1997a] to give a simpler approach to the subject. In the Chapter 7 we defined rich adapted spaces directly without introducing neocompact sets. In this chapter we will give a quick overview of the theory of neometric spaces. The neocompact sets will be defined here as countable intersections of basic sections, and their closure properties will be proved as theorems which hold in any rich adapted space.

8A. Neocompact sets

We assume throughout this section that $(\Omega, \mathcal{G}_t, P)$ is a rich $\mathbb{L}$-adapted space.

DEFINITION 8A.1.   *A* **neocompact set** *(in $\mathcal{M}$ over $(\Omega, \mathbb{L})$) is a countable intersection of basic sections.*

LEMMA 8A.2.   (*a*) *The family of neocompact sets in $\mathcal{M}$ is closed under finite unions and countable intersections.*

(*b*) *If A and B are neocompact sets in $\mathcal{M}$ and $\mathcal{N}$, then $A \times B$ is a neocompact set in $\mathcal{M} \times \mathcal{N}$.*

(*c*) *Every section of a neocompact relation is a neocompact set.*

(*d*) *Every compact set is neocompact (and is a basic section).*

PROOF.   We prove (d) and leave the rest as an exercise for the reader. Let C be a nonempty compact subset of $\mathcal{M}$. Choose a countable sequence $z_0, z_1, \ldots$ which is dense in C. Then z belongs to $\mathcal{N}$ where $N = M^{\mathbb{N}}$. Since *law* is continuous, the set $D = \{law(y, z) : y \in C\}$ is compact. We show that C is the basic section $C = \{y : law(y, z) \in D\}$. Let x belong to the right side. so that $law(x, z) = law(y, z)$ for some $y \in C$. Then some subsequence $\{z_m : m \in \mathbb{N}\}$ of z converges to y in $\mathcal{M}$. We have $law(x, z_m) = law(y, z_m)$ for all m. and

$$\lim_{m \to \infty} law(x, z_m) = law(x, y). \quad \lim_{m \to \infty} law(y, z_m) = law(y, y).$$

Thus $law(x, y) = law(y, y)$ and $x = y$. proving that $x \in C$.   ⊣

The following shows that neocompact sets act like compact sets in a rich $\mathbb{L}$-adapted space.

PROPOSITION 8A.3.   *For each Polish space M. the family of neocompact sets in $\mathbb{M}$ is countably compact.*

PROOF.   This follows easily from the fact that each neocompact set is a countable intersection of basic sections. and the family of basic sections is countably compact.   ⊣

We now prove two results from Keisler [1997a]. which show that the family of neocompact sets is closed under existential and universal projections. With these results. one can very quickly build up a large library of useful neocompact sets. See especially Fajardo and Keisler [1996a]. These theorems were called quantifier elimination theorems in Keisler [1997a]. because in that paper the neocompact sets were closed under projections by definition. so the theorem showed that the projections could be eliminated.

THEOREM 8A.4.   *For each neocompact relation A in $\mathcal{M} \times \mathcal{N}$. the* **existential projection**

$$E = \{x \in \mathcal{M} : (\exists y \in \mathcal{N})(x. y) \in A\}$$

is neocompact in $\mathcal{M}$.

PROOF.   (See Keisler [1998]. Theorem 5.1). Suppose first that A is a basic set. Let E be the existential projection of A. All the adapted functions involved in the definition of A are $\mathbb{F}$-adapted functions for some finite $\mathbb{F} \subseteq \mathbb{L}$. A is closed under convergence in $\mathbb{F}$-adapted distribution (i. e. closed under $\to_{\mathbb{F}}$-limits). We claim that E is closed under convergence in $\mathbb{F}$-adapted distribution. Let $x_n \Rightarrow_{\mathbb{F}} x$ with $x_n \in E$. Choose y_n such that $(x_n, y_n) \in A$. By Proposition 7B.8 (b), some subsequence of (x_n, y_n) converges in $\mathbb{F}$-adapted distribution, say to (x', y'). Then $(x', y') \in A$ and $x \equiv_{\mathbb{F}} x'$. By Theorem 7C.4, Ω is saturated as an $\mathbb{F}$-adapted space, so there exists y such that $(x, y) \equiv_{\mathbb{F}} (x', y')$. Therefore $x \in E$, and the claim is proved.

By Proposition 7A.1 (a), the set $\{law(x) : x \in E\}$ is contained in a compact set C. Choose a countable basis K of closed subsets of C and a countable $\equiv_{\mathbb{F}}$-dense set G of $\mathbb{F}$-adapted functions on $\mathcal{M}$. Let E_1 be the intersection of all basic sets in $\mathbb{M}$ which contain E and are of the form $\{x : law(x) \in C', g(x) \in D'\}$ such that $C' \in K$, $g \in G$, and D' is a finite union of closed intervals with rational endpoints. Then $E \subseteq E_1$, and E_1 is a countable intersection of basic sets, so E_1 is neocompact. Suppose $z \notin E$. Then there is a $\delta > 0$ and a finite set $G_0 \subseteq G$ such that E is disjoint from the set

$$J = \{x : d(law(x), law(z)) < \delta, (\forall g \in G_0)|E[g(x)] - E[g(z)]| < \delta\}.$$

We have $z \in J$, but the complement of J contains E_1. Therefore $z \notin E_1$. This proves that $E = E_1$, so E is neocompact.

The proof in the case that A is a basic section is similar but the parameter z must be carried along. Finally, suppose A is a countable intersection of basic sections A_n. By countable compactness we have

$$\{x : \exists y (x, y) \in A\} = \bigcap_n \{x : \exists y (x, y) \in A_n\}.$$

Therefore the projection $\{x : \exists y (x, y) \in A\}$ is a neocompact set. $\dashv$

We now prove closure under universal projections. This result needs the adapted version of the Skorokhod property.

THEOREM 8A.5. *For each neocompact relation A in $\mathcal{M} \times \mathcal{N}$ and nonempty basic set B in $\mathcal{N}$, the* **universal projection**

$$U = \{x \in \mathcal{M} : (\forall y \in B)(x, y) \in A\}$$

is neocompact in $\mathcal{M}$.

PROOF. (See Keisler [1998], Theorem 5.5). First let A be a basic set. Take a finite $\mathbb{F}$ such that both A and B are basic sets over $(\Omega, \mathbb{F})$. We claim that U is closed under convergence in $\mathbb{F}$-adapted distribution. Let $x_n \in U$ and $x_n \Rightarrow_{\mathbb{F}} x$. By the Skorokhod Property, Proposition 7B.8 (c), there exist $x'_n \equiv_K x_n$ such that x'_n converges to x in probability. Let $y \in B$. By the $\mathbb{F}$-adapted back and forth property, there exists $z \in B$ such that $(x_n, z) \equiv_{\mathbb{F}} (x'_n, y)$. Then $(x_n, z) \in A$, so $(x'_n, y) \in A$. The sequence (x'_n, y) converges in $\mathcal{M} \times \mathcal{N}$ to (x, y), so $(x, y) \in A$ and $x \in U$. This proves the claim. The rest of the proof is the same as the proof of Theorem 8A.4. $\dashv$

8B. Approximation theorem

We have been working with neocompact sets, which are analogous to compact sets. We now continue the theory by introducing the analogs of closed sets and continuous functions. We continue to assume throughout that Ω is a rich $\mathbb{L}$-adapted space.

DEFINITION 8B.1. (a) *A set $C \subseteq M$ is* **neoclosed** *in M if $C \cap D$ is neocompact in M for every neocompact set D in M.*

(b) *Let $D \subseteq M$. A function $f : D \to N$ is* **neocontinuous** *from M to N if for every neocompact set $A \subseteq D$ in M, the restriction $f|A = \{(x, f(x)) : x \in A\}$ of f to A is neocompact in $M \times N$.*

The following is a list of facts taken from Fajardo and Keisler [1996a]. They show that the new notions resemble quite closely the classical notions of compactness. closedness and continuity.

BASIC FACTS 8B.2. (a) *Every neocompact set is bounded and neoclosed.*

(b) *Every separable neocompact set is compact.*

(c) *Every neoclosed set is closed.*

(d) *Every neocontinuous function is continuous.*

(e) *If C is neocompact in M and $r \in \mathbb{R}$ then the set*

$$C^r = \{x \in M : \rho(x, C) \le r\}$$

is neoclosed in M. and $\rho(x, y) = r$ for some $y \in C$.

(f) *If $f : D \to N$ is neocontinuous from M to N and $A \subseteq D$ is neocompact in M, then the set $f(A) = \{f(x) : x \in A\}$ is neocompact in N.*

(g) *If $f : C \to N$ is neocontinuous from M to N, C is neoclosed in M, and D is neoclosed in N, then $f^{-1}(D) = \{x \in C : f(x) \in D\}$ is neoclosed in M.*

(h) *Compositions of neocontinuous functions are neocontinuous.*

PROOF. First note that for each M and real r. there are closed sets $A_r, B_r \subseteq Meas(M)$ such that for all $x, y \in M$,

$$\rho_0(x, y) \le r \text{ iff } law(x, y) \in A_r. \ \rho_0(x, y) \ge r \text{ iff } law(x, y) \in B_r.$$

Now prove (a), (b), and (e) using this fact. existential projection (Theorem 8A.4). and countable compactness (Proposition 8A.3). Parts (c) and (d) follow from the fact that every compact set is neocompact (Lemma 8A.2). Parts (f) – (h) follow from existential projection. ⊣

Constant functions, the identity function on M. and the projection functions on $M \times N$ are obviously neocontinuous. As of now we have not seen any nontrivial neocontinuous functions. The next theorem gives us plenty of them, and begins to connect things up with the previous chapters. We omit the proof since it requires a good deal of work (see Fajardo and Keisler [1996a]), but nonetheless we will use the theorem freely.

THEOREM 8B.3. *Each of the following functions is neocontinuous.*

(a) *The law function from M into $Meas(M)$.*

(b) *The distance function p_0 from $\mathcal{M} \times \mathcal{M}$ to $\mathcal{R}$.*

(c) *If f is continuous and the domain of f is separable then f is neocontinuous.*

(d) *If C is nonempty and basic in $\mathcal{M}$ then the function $f(x) = p(x, C)$ is neocontinuous from $\mathcal{M}$ to $\mathcal{R}$.*

(e) *If $f : M \to N$ is continuous, the function $\widehat{f} : \mathcal{M} \to \mathcal{N}$ defined by $\widehat{f}(x)(\omega) = f(x(\omega))$ is neocontinuous.*

(f) *If $h : \mathcal{M} \to \mathbb{R}$ is a bounded neocontinuous function then the function $E(h) : \mathcal{M} \to \mathbb{R}$ defined by $(E(h))(x) = E[h(x)]$ is neocontinuous.*

(g) *Each $\mathbb{L}$-adapted function is neocontinuous from $\mathcal{M}$ to $\mathcal{R}$.* ⊣

It is natural to ask for examples of sets which are neocompact but not basic, and functions which are continuous but not neocontinuous. There are always such examples ia a rich probability space.

EXAMPLE 8B.4. *Let $A_n, n \in \mathbb{N}$ be a countable independent family of sets of measure $1/2$. The set*

$$C = \{\mathbb{I}_B : P[B \Delta A_n] = 1/2 \text{ for all } n \in \mathbb{N}\}$$

is neocompact but not basic. The set

$$D = \{\mathbb{I}_B : p_0(\mathbb{I}_B, C) = 1/2\}$$

is closed but not neoclosed. The function $f(x) = p_0(x, C)$ is continuous but not neocontinuous.

PROOF. See Fajardo and Keisler [1996a], Example 4.17. Let $M = \{0, 1\}$. Then the space $Meas(M)$ is compact and $\mathcal{M}$ is basic. For each $k \in \mathbb{N}$ the set

$$C_k = \{\mathbb{I}_B : P[B \Delta A_n] = 1/2 \text{ for all } n < k\}.$$

is a basic section in $\mathcal{M}$, and $C = \bigcap_k C_k$, so C is neocompact. C is nonempty by countable compactness. D is clearly closed, and f is continuous because C is closed. Suppose D is neoclosed. Then $D \cap C_k$ is a decreasing chain of neocompact sets and $A_k \in D \cap C_k$, but $D \cap \bigcap_k C_k = \emptyset$. This contradicts countable compactness, so D cannot be neoclosed. We have $D = f^{-1}(\{1/2\})$, so by Basic Facts 8B.2 (c) and (g), f cannot be neocontinuous. By Basic Facts 8B.2 (c), C is not basic. ⊣

Another example of this kind is related to Example 7E.1, and is from Fajardo and Keisler [1996a], Example 5.7.

EXAMPLE 8B.5. *Let $\mathbb{L}$ contain a strictly decreasing sequence s_n converging to a number $t \in [0, 1)$. Let $\mathcal{F}_t$ be the P-completion of the σ-algebra $\bigcap\{\mathcal{G}_{s_n} : n \in \mathbb{N}\}$. The set $C = \{\mathbb{I}_A : A \in \mathcal{F}_t\}$ is neocompact but not basic. The set $D = \{\mathbb{I}_B : p_0(\mathbb{I}, C) = 1/2\}$ is closed but not neoclosed. The function $f(x) = p_0(x, C)$ is continuous but not neocontinuous .*

PROOF. The proof of Example 7E.1 shows that D is not neoclosed. The rest of the proof is like the preceding example. ⊣

To finish up this quick overview we prove a result which is crucial for the deeper applications, together with the Approximation Theorem that we mentioned in the introduction of this chapter. (See Fajardo and Keisler [1996a].)

THEOREM 8B.6. (*Diagonal Intersection Theorem*) *Let* M *be Polish, for each* $n \in \mathbb{N}$ *let* A_n *be a neocompact set, and let* $\varepsilon_n \to 0$. *Then*

$$A = \bigcap_n ((A_n)^{\varepsilon_n})$$

is a neocompact set, and if each finite intersection is nonempty then A *is nonempty.*

PROOF. By Basic Fact , for each n the set $(A_n)^{\varepsilon_n}$ is neoclosed. Therefore A is neoclosed. Since A_n is neocompact and *law* is neocontinuous, the image *law* (A_n) is neocompact by Basic Fact 8B.2 (f). Then by Basic Fact 8B.2 (b), *law* (A_n) is compact. The Prohorov metric d on *Meas*(M) has the property that

$$d(law(x), law(y)) \leq \rho_0(x, y).$$

Therefore for each n, *law* $((A_n)^{\varepsilon_n}) \subseteq (C_n)^{\varepsilon_n}$. and hence *law* $(A) \subseteq C$ where $C = \bigcap_n((C_n)^{\varepsilon_n})$. Since $\varepsilon_n \to 0$ and each C_n is totally bounded, the set C is totally bounded. Since C is also closed, it is compact. Therefore A is contained in the basic neocompact set *law* $^{-1}(C)$ in M. and hence A is neocompact in M.

We leave as an exercise the proof that A is nonempty if each finite intersection is nonempty. ⊣

The Diagonal Intersection Theorem implies other closure properties for the neocompact sets. Here is one such example that will be used in the proof of the Approximation Theorem and at the same time will give us an idea of what proofs in the theory of neometric spaces look like.

LEMMA 8B.7. *Let* A *be a neocompact set and let* (x_n) *be a sequence in* M *such that* $\lim_{n\to\infty} \rho(x_n, A) = 0$. *Then the set*

$$B = A \cup \{x_n : n \in \mathbb{N}\}$$

is neocompact.

PROOF. Choose a decreasing sequence (ε_n) such that $\varepsilon_n \geq \rho(x_n, A)$ and $\varepsilon_n \to 0$ as $n \to \infty$. Let $A_n = A \cup \{x_m : m \leq n\}$. Then each A_n is neocompact, and $B \subseteq \bigcap_n(A_n)^{\varepsilon_n}$. We also have the opposite inclusion $B \supseteq \bigcap_n(A_n)^{\varepsilon_n}$, because if $y \notin B$ then $y \notin \{x_n : n \in \mathbb{N}\}$, and $\rho(y, A) > 0$ by Basic Fact 8B.2 (e), so $y \notin (A_n)^{\varepsilon_n}$ for some n. By the Diagonal Intersection Theorem, B is neocompact in M. ⊣

THEOREM 8B.8. (*Approximation Theorem*) *Let* A *be neoclosed in* M *and* $f :$ $A \to M$ *be neocontinuous from* M *into* N. *Let* B *be neocompact in* M *and* D *neoclosed in* N. *If*

$$(\exists x \in A \cap B^r) f(x) \in D^r$$

for each real $r > 0$, *then*

$$(\exists x \in A \cap B) f(x) \in D.$$

PROOF. By hypothesis there is a sequence (x_n) in A such that for each n, $x_n \in B^{1/n}$ and $f(x_n) \in D^{1/n}$. Let $C = A \cap B$. By the preceding lemma, for each m the set

$$C_m = A \cap (B \cup \{x_n : m \le n \in \mathbb{N}\}) = C \cup \{x_n : m \le n \in \mathbb{N}\}$$

is a neocompact subset of A. Since f is neocontinuous on A, the graph

$$G_m = \{(x, f(x)) : x \in C_m\}$$

is neocompact in $\mathcal{M} \times \mathcal{N}$.

Since f is neocontinuous and C_1 is neocompact in $\mathcal{M}$, the set $f(C_1)$ is neocompact in $\mathcal{N}$. We may choose $y_n \in D$ such that $\rho(f(x_n), y_n) \le 1/n$. Then $y_n \in (f(C_1))^{1/n}$. Using the preceding lemma again, the set

$$E = D \cap [f(C_1) \cup \{y_n : n \in \mathbb{N}\}]$$

is neocompact in $\mathcal{N}$. We have $y_m \in E$ and $f(x_m) \in E^{1/m}$ for each m. By Basic Fact 8B.2 (e), $E^{1/m}$ is neoclosed, so

$$H_m = G_m \cap (A \times E^{1/m})$$

is a decreasing chain of neocompact sets. Since $(x_m, f(x_m)) \in H_m$, H_m is nonempty. By countable compactness there exists $(x, z) \in \bigcap_m H_m$. Then $x \in C, z \in D$, and $z = f(x)$ as required. ⊣

REMARK 8B.9. *It is an easy exercise to prove the analog of the Approximation Theorem with compact, closed, and continuous in place of neocompact, neoclosed, and neocontinuous.*

8C. Some applications

In this section we give some applications which illustrate how rich adapted spaces and the Approximation Theorem can be used in stochastic analysis.

Throughout this section we let Ω be a rich $\mathbb{B}$-adapted space with the filtration $\mathcal{G}_s, s \in \mathbb{B}$, and let Ω^{rt} be the corresponding right continuous adapted space with the filtration $\mathcal{F}_t = \bigcap_{s>t} \mathcal{G}_s$. All neometric concepts, such as neocompact or neocontinuous, will be meant over $(\Omega, \mathbb{B})$. But adapted probability concepts, such as martingales, stopping times, or stochastic integrals, will be on Ω^{rt} (i.e., with respect to the right continuous filtration $\mathcal{F}_t$).

To apply the Approximation Theorem one needs to know that the objects involved have the required neometric properties. Many of the fundamental concepts in stochastic analysis have these properties. Here are some examples from Fajardo and Keisler [1996a]. The proofs make repeated use of the closure properties of neocompact sets.

THEOREM 8C.1. *Let M be Polish.*
(a) The set of $\mathcal{F}_t$-Brownian motions is neocompact.
(b) The set of $\mathcal{F}_t$-stopping times is neocompact.
(c) The set of $\mathcal{F}_t$-adapted processes with values in M is neoclosed.

(*d*) *The set of continuous $\mathcal{F}_t$-adapted processes with values in M is neoclosed.*

(*e*) *Each m-fold conditional process $f \in CP$ on Ω^{rt} is a neocontinuous function from $\mathcal{M}$ into $L^0(\Omega \times [0, 1]^m, \mathbb{R})$, and $E[f]$ is neocontinuous from $\mathcal{M}$ into $L^0([0, 1]^m, \mathbb{R})$.*   ⊣

Many applications are existence theorems stating that a set is nonempty with the additional bonus that the set is neocompact. The advantage of knowing that the set of solutions is neocompact is that one can combine two or more existence theorems to get a simultaneous solution.

Whenever we have a neocompact set and a real-valued neocontinuous function, we get an optimization theorem for free. because the image of the neocontinuous function is compact and thus has a maximum and minimum. Here are some examples using Theorem 8C.1.

COROLLARY 8C.2. (*a*) *For any continuous real-valued stochastic process x, the set of Brownian motions at minimum distance from x in $L^0(\Omega, C([0, 1]))$ is a nonempty neocompact set.*

(*b*) *For any random variable x and one-fold conditional process $f \in CP$ on Ω^{rt}, the set of Brownian motions b such that $\int E[f(b, x)]$ is a minimum is a nonempty neocompact set.*

(*c*) *For any random variable $x \in L^0(\Omega, [0, 1])$ the set of stopping times at minimum distance from x in $L^0(\Omega, [0, 1])$ is a nonempty neocompact set.*   ⊣

Here are some applications of the Approximation Theorem from Fajardo and Keisler [1996a] and Keisler [1997b].

THEOREM 8C.3. *Let $w(\omega, s)$ be a continuous martingale in $\mathcal{R}^d$ and let g be a uniformly bounded adapted process with values in the space $C(\mathbb{R}^d, \mathbb{R}^{d \times d})$. Then the set of continuous martingales x in $\mathcal{R}^d$ such that*

$$(21) \qquad x(\omega, t) = \int_0^t g(\omega, s)(x(\omega, s))dw(\omega, s).$$

is nonempty and neocompact.

PROOF. We give the main ideas. Let A be the neoclosed set of pairs (u, x) where $u \in [0, 1]$ and x is an adapted process with values in $\mathbb{R}^d$. Using the convention that $x(\omega, t) = 0$ for $t < 0$. show that the function

$$h(u, x)(t) = \int_0^t g(\omega, s)(x(\omega, s - u))dw(\omega, s)$$

is neocontinuous from A into the set of continuous adapted processes with values in $\mathbb{R}^d$. Since g is uniformly bounded. one can show that the range of h is contained in a neocompact set C of continuous adapted processes. Let B be the neocompact set $\{0\} \times C$. and let f be the neocontinuous function $f(u, x) = \rho^0(x, h(u, x))$. One can easily construct approximate solutions with a small time delay to show that for each $\delta > 0$ there exists $(u, x) \in A$ with $u \leq \delta$ and $f(u, x) \leq \delta$. Since $u \leq \delta$ and $h(u, x) \in C$ we have $(u, x) \in A \cap B^\delta$ and $f(u, x) \in \{0\}^\delta$. Therefore by the Approximation Theorem there exists x such that $(0, x) \in A \cap B$ and

$f(0, x) = 0$, which is the required condition. The set of such x is neocompact because A is neoclosed, B is neocompact, and f is neocontinuous.   ⊣

The following results use the Approximation Theorem in a similar way, so we will leave the proofs to the reader.

THEOREM 8C.4. *Let w be a continuous martingale, and let*

$$f : C([0, 1], \mathbb{R}^d) \to \mathbb{R}$$

be a bounded continuous function. Then the set of solutions x of equation (21) such that $E[f(x(\omega))]$ is a minimum is nonempty and neocompact.   ⊣

THEOREM 8C.5. *Let B be a nonempty neocompact set of continuous martingales, and let*

$$f : C([0, 1], \mathbb{R}^d \times \mathbb{R}^d) \to \mathbb{R}$$

be a bounded continuous function. Then the set of pairs (x, w) such that $w \in B$, (x, w) solves equation (21), and $E[f(x(\omega), w(\omega))]$ is a minimum, is nonempty and neocompact.   ⊣

THEOREM 8C.6. *Suppose that we have a sequence of equations*

$$x(\omega, t) = \int_0^t g_n(\omega, s)(x(\omega, s)) dw_n(\omega, s)$$

where each g_n is a bounded adapted process with values in the space $C(\mathbb{R}^d, \mathbb{R}^{d \times d}))$, and w_n is a continuous martingale with values in $\mathbb{R}^d$. Assume that for each n there exists an x which is a solution of the first n equations. Then there exists an x which is a simultaneous solution of all the equations, and the set of all such x is again neocompact.   ⊣

We conclude this chapter by keeping a promise made in Chapter 7, to prove the following improvement of Theorem 7E.3. The proof was postponed until this point because it uses the fact that conditional processes are neocontinuous, Theorem 8C.1 (e).

THEOREM 8C.7. *For every rich $\mathbb{B}$-adapted space Ω, the corresponding right continuous adapted space Ω^{rt} is saturated.*

PROOF. As usual, it suffices to show that Ω^{rt} is universal and has the back and forth property for random variables. Let x' be an M-valued random variable on some right continuous adapted space. By Proposition 7E.2 (a), there is a sequence x_n of random variables on Ω^{rt} which converges in adapted distribution to x'. Then the set

$$C = \{law(x_n) : n \in \mathbb{N}\} \cup \{law(x')\}$$

is compact, and hence the set

$$D = \{x : law(x) \in C\}$$

is neocompact (and even basic). By Proposition 3D.2 there is a countable dense set $\{h_k : k \in \mathbb{N}\}$ of conditional processes. Let h_k be $m(k)$-fold. Let d be the metric for $Meas(M)$ and d_k the metric for $L^0([0,1]^{m(k)}, \mathbb{R})$. Consider the sets

$$A_n = \left\{ x \in \mathcal{M} : d(law(x), law(x')) \leq \frac{1}{n} \right\},$$

$$B_{kn} = \left\{ x \in \mathcal{M} : d_k(E[f_k(x)], E[f_k(x')]) \leq \frac{1}{n} \right\}.$$

The functions d and d_k are neocontinuous by Theorem 8B.3 (c). By Theorem 8B.3 (a) and closure under composition, the function $x \mapsto d(law(x), law(x'))$ is neocontinuous. By Theorem 8C.1 (e), for each k the function $x \mapsto d_k(E[f_k(x)], E[f_k(x')])$ is neocontinuous. Then by Basic Facts 8B.2 (g), each of the sets A_n and B_{kn} is neoclosed, and hence meets D in a neocompact set. Moreover, since $x_n \Rightarrow x'$, D meets each intersection of finitely many of the sets A_n and B_{kn}. By countable compactness there exists an $x \in D$ which belongs to A_n and B_{kn} for every k, n. Then $x \equiv x'$, and Ω^{rt} is universal.

The back and forth property for Ω^{rt} is proved by a similar argument using Proposition 7E.2 (b).   ⊣

CHAPTER 9

ENLARGING SATURATED SPACES

In this chapter we turn to a question that is in the spirit of Chapter 6: *If we start with a saturated adapted space and enlarge the filtration, when will the new space be saturated?* We have postponed this topic until now because we need the result from Chapter 7 that Loeb adapted spaces are saturated. We will prove a result that gives another illustration of the usefulness of adapted Loeb spaces.

This chapter depends on Chapter 7 but is independent of Chapter 8. We will only consider right continuous adapted spaces, and will never use the $\mathbb{L}$-adapted spaces introduced in Chapters 7 and 8.

9A. Atomic enlargments

Barlow and Perkins [1989] and Barlow and Perkins [1990] made use of adapted distributions and adapted Loeb spaces to study properties of solutions of stochastic differential equations. In his paper "Enlarging saturated filtrations", Hoover [1990] extended their results in a way which fits naturally into our general model theoretic setting. As usual, his solution led to a more general construction. Here we just single out one aspect of this work where a result on hyperfinite adapted spaces is applied to a problem about arbitrary adapted spaces. This gives a good illustration the idea behind the "Come-back Problem" in Fremlin [1989].

We state the situation in the broadest possible terms. Suppose we are given adapted spaces $\Omega = (\Omega, \mathcal{F}_t, P)$ and $\Gamma = (\Omega, \mathcal{G}_t, P)$ such that $\mathcal{F}_t \subseteq \mathcal{G}_t \subseteq \mathcal{F}_1$ for each t (that is, Γ is obtained from Ω by adding sets from $\mathcal{F}_1$ to the filtration). Assume that Ω is saturated. When is Γ saturated?

We have been intuitively thinking of saturated spaces as spaces which have plenty of measurable sets. With this idea in mind, one might guess that if Ω is saturated then every Γ such that $\mathcal{F}_t \subseteq \mathcal{G}_t \subseteq \mathcal{F}_1$ for each t is saturated. But this guess would be wrong; A counterexample is given in Hoover [1990]. We will see in this chapter that saturation will be preserved if only a few measurable sets are added to the filtration.

The result of Barlow and Perkins involves the **optional** σ-algebra for an adapted space Ω, which is defined as the smallest σ-algebra of subsets of $\Omega \times [0, 1]$ for which each cadlag adapted stochastic processes on Ω is measurable. A member of the optional σ-algebra is called an **optional set**. Given a function $L : \Omega \to [0, 1]$, enlarge the filtration $(\mathcal{F}_t)$ to the smallest filtration $(\mathcal{F}_t^L)$ so that L becomes an $(\mathcal{F}_t^L)$-stopping time. Then ask whether the space $\Omega^L = (\Omega, \mathcal{F}_t^L, P)$ is saturated. Barlow and Perkins [1989] proved the following result.

THEOREM 9A.1. *Let Ω be an atomless Loeb adapted space, x be a process on Ω, O be an optional set for Ω, and $L : \Omega \to \mathbb{R}$ be the end of O, that is, the function*

$$L(\omega) = \sup(\{0\} \cup \{t \in [0, 1] : (\omega, t) \in O\}).$$

Then Ω^L is an atomless Loeb adapted space (and hence is saturated by Corollary 7E.4).   ⊣

In general, it is clear that L is not an $(\mathcal{F}_t)$-stopping time because the value of $\min(t, L(\omega))$ depends on the future of ω after time t. It is easy to see that for each t, $\mathcal{F}_t^L$ is the σ-algebra generated by $\mathcal{F}_t$ and the single set $\{\omega : L(\omega) \le t\}$.

In Theorem 9A.1, the adapted space Ω^L has the same measurable sets and probability measure as Ω. It is an adapted space of the form $(\Omega, \mathcal{G}_t, P)$ where $\mathcal{F}_t \subseteq \mathcal{G}_t \subseteq \mathcal{F}_1$ for each t. The improvement of Theorem 9A.1 in Hoover [1990] applies to extensions of Ω of this kind such that only a few sets are added to the filtration.

DEFINITION 9A.2. *By an **enlargement** of an adapted space $\Omega = (\Omega, \mathcal{F}_t, P)$ we will mean an adapted space $\Gamma = (\Omega, \mathcal{G}_t, P)$ such that $\mathcal{F}_t \subseteq \mathcal{G}_t \subseteq \mathcal{F}_1$ for each t (or equivalently, $\Omega \sqsubseteq \Gamma$ and $\mathcal{G}_1 = \mathcal{F}_1$.)*

We say that a σ-algebra $\mathcal{G}$ is atomic over a σ-algebra $\mathcal{F}$ if $\mathcal{G} = \mathcal{F} \vee \mathcal{H}$ where $\mathcal{H}$ is generated by countably many disjoint sets.

*An **atomic enlargement** of Ω is an enlargement Γ of Ω such that $\mathcal{G}_t$ is atomic over $\mathcal{F}_t$ for every t in a countable dense set $D \subseteq [0, 1]$.*

Note that in Theorem 9A.1, Ω^L is an atomic enlargement of Ω. Our goal, which will be achieved in the next section, is to prove the following result of Hoover [1990].

If Ω is a saturated adapted space and Γ is an atomic enlargement of Ω, then Γ is also saturated.

Roughly, the idea will be to first prove that Γ is atomless, then prove that Γ is a Loeb adapted space if Ω is, and then use a model theoretic argument to prove the theorem for all adapted spaces.

Hoover's proof used the notion of convergence in adapted distribution, which we have not defined here. Instead we will use the fact that all atomless adapted Loeb spaces are saturated. (This is the reason that we postponed this section until now).

To prove that Γ is atomless, we need three lemmas about extensions of σ-algebras. We leave the proofs of the first two of these lemmas as exercises for the reader.

LEMMA 9A.3. *If $\mathcal{F}' \subseteq \mathcal{F} \subseteq \mathcal{G} \subseteq \mathcal{G}'$ are σ-algebras over Ω and $\mathcal{G}$ is atomless over $\mathcal{F}$, then $\mathcal{G}'$ is atomless over $\mathcal{F}'$.*   ⊣

LEMMA 9A.4. *Let $\mathcal{F}$ be a $\sigma-$algebra over Ω, $H = (H_n)$ be a countable measurable partition of Ω, $\mathcal{G} = \mathcal{F} \vee H$, and x be a bounded random variable. Then*

$$E[x|\mathcal{G}] = \sum_n \frac{E[\mathbb{I}_{H_n} x|\mathcal{F}]}{P[H_n|\mathcal{F}]} \cdot \mathbb{I}_{H_n}.$$
⊣

LEMMA 9A.5. *If $\mathcal{F} \subseteq \mathcal{G} \subseteq \mathcal{H}$ are σ-algebras over Ω, $\mathcal{H}$ is atomless over $\mathcal{F}$, and $\mathcal{G}$ is atomic over $\mathcal{F}$, then $\mathcal{H}$ is atomless over $\mathcal{G}$.*

PROOF. We have $\mathcal{G} = \mathcal{F} \vee H$ for some countable measurable partition H of Ω. Consider a set $B \subseteq H_m$ of positive measure where $B \in \mathcal{H}$. Since $\mathcal{H}$ is atomless over $\mathcal{F}$, there is a set $C \subseteq B$ in $\mathcal{H}$ such that $0 < P[C|\mathcal{F}] < P[B|\mathcal{F}]$ on a set of positive measure.

Taking $x = \mathbb{I}_C$ in the preceding lemma, we have

$$P[C|\mathcal{G}] = \sum_n \frac{P[C \cap H_n|\mathcal{F}]}{P[H_n|\mathcal{F}]} \cdot \mathbb{I}_{H_n} = \frac{P[C|\mathcal{F}]}{P[H_m|\mathcal{F}]} \cdot \mathbb{I}_{H_m} \ a.s.$$

Similarly,

$$P[B|\mathcal{G}] = \frac{P[B|\mathcal{F}]}{P[H_m|\mathcal{F}]} \cdot \mathbb{I}_{H_m} \ a.s.$$

Therefore $0 < P[C|\mathcal{G}] < P[B|\mathcal{G}]$ on a set of positive measure.   ⊣

It is now easy to show that an atomic enlargement Γ is atomless.

PROPOSITION 9A.6. *If Ω is an atomless adapted space and Γ is an atomic enlargement of Ω, then Γ is also atomless.*

PROOF. Since $\mathcal{F}_0$ is atomless, $\mathcal{G}_0$ is atomless. Let D be a dense subset of $[0, 1]$ such that $\mathcal{G}_t$ is atomic over $\mathcal{F}_t$ for all $t \in D$. Let $s < t$ in $[0, 1]$. Take $u, v \in D$ with $s < u < v < t$. By Lemmas 9A.3 and 9A.5, $\mathcal{G}_v$ is atomless over $\mathcal{G}_u$. Then using Lemma 9A.3 again, $\mathcal{G}_t$ is atomless over $\mathcal{G}_s$.   ⊣

The next step is to show that an atomic enlargement of a Loeb adapted space is again a Loeb adapted space.

THEOREM 9A.7. *Let $\Omega = (\Omega, \mathcal{F}_t, P)$ be a Loeb adapted space, and let $\Gamma = (\Gamma, \mathcal{G}_t, P)$ be an atomic enlargement of Ω. Then Γ is a Loeb adapted space.*

PROOF. For simplicity we take the underlying internal probability space $(\Omega, \mathcal{A}_1, P)$ so that Ω is a hyperfinite set and $\mathcal{A}_1$ is the set of all internal subsets of Ω. We prove the particular case of Theorem 9A.7 where the atomic enlargement is obtained by adding the same partition at all times (i.e., for every t, $\mathcal{G}_t = \mathcal{F}_t \vee \sigma(H)$, where H is a countable Loeb measurable partition of Ω). The general case will be left as an exercise for the reader (see Remolina [1993] and Fremlin [1989]).

Let $(\Omega, (\mathcal{A}_t)_{t \in \mathbb{T}}, \bar{P})$ be an internal adapted space, and let the corresponding Loeb adapted space be $(\Omega, \mathcal{F}_t, P)$, where $\mathbb{T}$ is the hyperfinite time line. We do not assume that Ω is the internal hyperfinite adapted space constructed in Chapter 2. (Even if we did start with the internal hyperfinite adapted space, we would end up with another adapted Loeb space at the end). All we have to do is to find an internal filtration $(\mathcal{B}_t)_{t \in \mathbb{T}}$ such that $(\mathcal{G}_t)_{t \in [0,1]}$ is the filtration of the corresponding Loeb adapted space.

The idea is natural. For each H_n in H find an internal $\widehat{H}_n$ such that $P[\widehat{H}_n \bigtriangleup H_n] = 0$. By induction on n, we can also arrange it so that the sets $\widehat{H}_n$ are pairwise disjoint. Then, modulo a set of measure zero, the $\widehat{H}_n$'s form a Loeb measurable partition of Ω. By saturation we can extend the sequence $(\widehat{H}_n)_{n \in \mathbb{N}}$ to a hyperfinite partition $\widehat{H} = (\widehat{H}_n)_{n \leq J}$ of Ω. Notice that it really doesn't matter which J we choose as long as it is infinite. Then $\widehat{H}$ is internal and clearly the $\widehat{H}_i$'s that correspond to infinite i's all have infinitesimal probability.

For each $t \in \mathbb{T}$, let $\mathcal{B}_t$ be the *algebra of sets generated by $\mathcal{A}_t \cup \widehat{H}$. Then $(\mathcal{B}_t)_{t \in \mathbb{T}}$ is an internal *filtration. It is an easy exercise to show that the filtration for the corresponding Loeb adapted space is precisely $(\mathcal{G}_t)$. We leave it to the reader. ⊣

9B. Saturated enlargements

We now turn to arbitrary saturated adapted spaces. As we have explained in the last section, our goal is to prove the following theorem from Hoover [1990].

THEOREM 9B.1. *Let $\Gamma = (\Gamma, \mathcal{G}_t, Q)$ be a saturated adapted space, and let $\Gamma' = (\Gamma, \mathcal{G}'_t, Q)$ be an atomic enlargement of Γ. Then Γ' is saturated.*

PROOF. Again, we will just give the proof in the simple case that $\mathcal{G}'_t = \mathcal{G}_t \vee K$ for all t, where $K = (K_n)_{n \in \mathbb{N}}$ is a countable measurable partition of Γ. Let $\Omega = (\Omega, \mathcal{F}_t, P)$ be our usual hyperfinite adapted space (any atomless hyperfinite Loeb adapted space will do). Consider the expanded structure $(\Gamma, \mathbb{I}_{K_n})_{n \in \mathbb{N}}$. By saturation of Ω there is a countable sequence of Loeb measurable sets H_n such that

(22) $$(\Gamma, \mathbb{I}_{K_n})_{n \in \mathbb{N}} \equiv (\Omega, \mathbb{I}_{H_n})_{n \in \mathbb{N}}.$$

Then, of course, the H_n's form a partition of Ω (again, modulo a measure zero set). Let $\Omega' = (\Omega, \mathcal{F}'_t, P)$ be the adapted space such that $\mathcal{F}'_t = \mathcal{F}_t \vee H$ where $H = (H_n)_{n \in \mathbb{N}}$. Ω' is an atomless Loeb adapted space by Proposition 9A.6 and Theorem 9A.7, so Ω' is saturated by Corollary 7E.4.

Now let Λ be an arbitrary adapted space. Suppose x and x' are stochastic processes on Λ and Γ' respectively such that $(\Lambda, x) \equiv (\Gamma', x')$. Let y be another process on Λ. We want to find a process y' on Γ' such that $(\Lambda, x, y) \equiv (\Gamma', x', y')$. Let us take a detour.

First of all, since $\mathcal{G}'_1 = \mathcal{G}_1$, the process x' is also a process on Γ. From (22) and the saturation of Ω we can find a process z on Ω such that

(23) $$(\Gamma, \mathbb{I}_{K_n}, x')_{n \in \mathbb{N}} \equiv (\Omega, \mathbb{I}_{H_n}, z)_{n \in \mathbb{N}}.$$

We claim that

(24) $$(\Lambda, x) \equiv (\Gamma', x') \equiv (\Omega', z).$$

This follows easily from Lemma 9A.4. For every adapted function f we get that

$$E[f(x')|\mathcal{G}_t \vee K] = \sum_0^\infty \frac{E[\mathbb{I}_{K_n} f(x')|\mathcal{G}_t]}{Q[\mathbb{I}_{K_n}|\mathcal{G}_t]} \mathbb{I}_{K_n}.$$

and also the analogous statement in Ω with $f(z)$ in place of $f(x')$. This fact, together with (23), gives us (24).

We are ready to finish the proof. Using the saturation of Ω' we can find a process u on Ω' such that

$$(25) \qquad\qquad (\Lambda, x, y) \equiv (\Omega', z, u).$$

Going back to (23) and using the saturation of Γ, we can find a process y' on Γ such that

$$(26) \qquad\qquad (\Gamma, \mathbb{I}_{K_n}, x', y')_{n \in \mathbb{N}} \equiv (\Omega, \mathbb{I}_{H_n}, z, u)_{n \in \mathbb{N}}.$$

A similar argument to that given for (24) now gives us

$$(27) \qquad\qquad (\Gamma', x', y') \equiv (\Omega', z, u).$$

We now conclude from (25) and (27) that Γ' is saturated.   ⊣

The proofs of the last two theorems can be carried out in a natural way in the general case where the filtration is extended by adding a countable measurable partition at each time in a countable dense set D. The argument should be very similar to the two above. However, some care is needed in choosing the internal approximations of the partitions in the proof of Theorem 9A.7. It should be done in such a way that the internal sequence $\mathcal{B}_t$ which is obtained in the natural way is increasing with $t \in \mathbb{T}$.

Hoover [1990], using a stochastic analysis argument, gave an example of a saturated adapted space Ω with a non-saturated enlargement Γ. In fact, he showed even more. The next definition will help us describe his example.

DEFINITION 9B.2. Let $\Omega = (\Omega, \mathcal{F}_t, P)$ be an adapted space and x a cadlag stochastic process on Ω. Then $\mathcal{F}_t^x$ denotes the new filtration defined by $\mathcal{F}_t^x = \mathcal{F}_t \vee \sigma(x_s : s \leq t)$, and $\Omega^x = (\Omega, \mathcal{F}_t^x, P)$ is the adapted space with this new filtration.

Note that Ω^x is the smallest enlargement Γ of Ω such that x is Γ-adapted. Moreover, Ω^x has the property that for each time t, $\mathcal{F}_t^x$ is generated by $\mathcal{F}_t$ and a countable family of sets.

PROPOSITION 9B.3. Let Ω be a saturated adapted space.
(a) Ω has an enlargement of the form Ω^x which is not saturated.
(b) Ω has an elementary extension which is not saturated.

PROOF. For (a), take an $\mathcal{F}_1$-measurable Brownian motion x which is independent of $\mathcal{F}_t$ for each $t < 1$, and let $\Gamma = \Omega^x$. By a result of Billingsley [1979], there is a stochastic differential equation with continuous coefficients which does not have a solution which is measurable with respect to x. Use this to show that Ω^x is not saturated.

For part (b), let Υ be an adapted space whose filtration $\mathcal{E}_t$ is generated by a Brownian motion and take Λ to be the independent product $\Omega \times \Upsilon$. Λ fails to be saturated by the argument for part (a). and $\Omega \prec \Lambda$ by Lemma 6C.2. $\dashv$

It is natural to ask whether saturation is preserved in the downward direction. If Ω has a saturated enlargement, must Ω be saturated? The answer is "No;" Ω does not even have to be atomless. To get a counterexample, take Ω so that $\mathcal{F}_t$ is finite for $t < 1$ and $\mathcal{F}_1 = \mathcal{G}_1$. For an atomless example, let x be a Γ-adapted Brownian motion and form Ω by setting $\mathcal{F}_t = \sigma(x_s : s \le t)$ for $t < 1$ and again put $\mathcal{F}_1 = \mathcal{G}_1$.

If we put additional restrictions on Ω. we end up with another open problem.

PROBLEM 9B.4. *If Ω is atomless and has a saturated atomic enlargement, must Ω be saturated?*

If Ω is atomless and has a saturated enlargement of the form Ω^x for some cadlag process x on Ω. must Ω be saturated?

We conclude this chapter with one more result of Hoover [1990] which is related to the problem of enlarging filtrations and preserving saturation. We prove a lemma which does most of the work and then state the theorem.

LEMMA 9B.5. *For any adapted spaces Γ and Λ, if $(\Lambda, x, u) \equiv (\Gamma. y, v)$. then $(\Lambda^x, u) \equiv (\Gamma^y. v)$.*

PROOF. We use the Intrinsic Isomorphism Theorem 5A.10. Suppose that $h : (\Lambda, \mathcal{I}^{xu}) \simeq (\Gamma, \mathcal{I}^{yv})$ where $\mathcal{I}^{xu}$ is the intrinsic filtration of the pair $(x. u)$ with respect to Λ and similarly for Γ. All that is needed is to observe that h induces in the obvious way an adapted isomorphism $h' : (\Lambda^x, \mathcal{J}^u) \simeq (\Gamma^y, \mathcal{J}^v)$ where $\mathcal{J}^u$ is the intrinsic filtration of the process u with respect to Λ^x and similarly for Γ. (Hoover [1990] gave a different argument.) $\dashv$

THEOREM 9B.6. *Let Λ and Γ be saturated adapted spaces with stochastic processes x and y such that $(\Lambda, x) \equiv (\Gamma, y)$. If the adapted space Λ^x is saturated then the adapted space Γ^y is also saturated.*

PROOF. It is an easy exercise to prove this theorem from the preceding lemma, and we leave it to you. $\dashv$

Here is a problem that seems very natural after the above theorem.

PROBLEM 9B.7. *Given a saturated adapted space Γ. for which processes y on Γ is Γ^y saturated?*

As one can see, there are many places within the theory of stochastic processes where ideas and methods coming from logic can be useful. In fact. when browsing through probability journals, one gets the feeling that many results and ideas can be phrased and redone very nicely using nonstandard analysis and the model theory of stochastic processes.

REFERENCES

S. ALBEVERIO, J. FENSTAD, R. HOEGH-KROHN, AND T. LINDSTRØM
[1986] *Nonstandard methods in stochastic analysis and mathematical physics*, Academic Press.

DAVID ALDOUS
[1981] *Weak convergence and the general theory of processes*, preprint.

ROBERT ANDERSON
[1976] *A nonstandard representation for brownian motion and itô integration*, *Israel Journal of Mathematics*, vol. 25, pp. 15–46.

ROBERT ASH
[1972] *Real analysis and probability*, Academic Press.

MARTIN BARLOW AND EDWARD PERKINS
[1989] *Sample path properties of stochastic integrals and stochastic differentiation*, *Stochastics and Stochastic reports*, vol. 27, pp. 261–293.
[1990] *On pathwise uniqueness and expansion of filtrations*, *Seminaire de probabilities xxvi* (A. Azema et. al, editor), Lecture Notes in Mathemtics, no. 1426, Springer-Verlag, pp. 194–209.

PATRICK BILLINGSLEY
[1979] *Probability and measure*, Wiley.

C.C. CHANG AND H. JEROME KEISLER
[1990] *Model theory, third edition*, North Holland.

NIGEL CUTLAND
[1983] *Nonstandard measure theory and its applications*, *Bulletin of the London Mathematical Society*, vol. 15, pp. 529–589.

ROBERT DALANG
[1989] *Optimal stopping of two parameter processes on nonstandard probability spaces*, *Transactions of the American Mathematical Society*, vol. 373, pp. 697–719.

CLAUDE DELLACHERIE AND PAUL A. MEYER
[1978] *Probabilities and potential*, North-Holland.
[1988] *Probabilities and potential c*, North-Holland.

R. DURRETT
[1984] *Brownian motion and martingales in analysis*, Wadsworth.

STEVE ETHIER AND TOM KURTZ
[1986] *Markov processes*, Wiley.

SERGIO FAJARDO
[1985a] *Completeness theorems for the general theory of processes*, **Methods in mathematical logic** (Carlos DiPrisco, editor), Lecture Notes in Mathematics, no. 1130, Springer-Verlag, pp. 174–194.
[1985b] *Probability logic with conditional expectation*, **Annals of Pure and Applied Logic**, vol. 28, pp. 137–161.
[1987] *Intrinsic stochastic processes*, **Revista Colombiana de Matemáticas**, vol. 21, pp. 317–336.
[1990a] *Elementary embeddings and games in adapted probability logic*, **Archive for Mathematical Logic**, pp. 49–58.
[1990b] *Introducción al análisis no-estándar y sus aplicaciones en probabilidad*, Cuadernos de Probabilidad y Estadística Matemática, no. 3, Fondo Editorial Acta Científica Venezolana, Caracas.

SERGIO FAJARDO AND H. JEROME KEISLER
[1995] *Long sequences and neocompact sets*, **Developments in nonstandard analysis** (Nigel Cutland et al., editor), Longman, pp. 251–260.
[1996a] *Existence theorems in probability theory*, **Advances in Mathematics**, vol. 120, pp. 191–257.
[1996b] *Neometric spaces*, **Advances in Mathematics**, vol. 118, pp. 134–175.

SERGIO FAJARDO AND JAVIER PEÑA
[1997] *Games on probability spaces*, preprint.

DAVID FREMLIN
[1989] *Measure algebra*, **The handbook of boolean algebra, volume 3** (J.D. Monk and R. Bonnet, editors), North-Holland.

C. WARD HENSON AND H. JEROME KEISLER
[1986] *On the strength of nonstandard analysis*, **The Journal of Symbolic Logic**, vol. 51, pp. 377–386.

WILFRID HODGES
[1985] *Building models by games*, Student Texts 2, London Mathematical Society.

DOUGLAS HOOVER

[1984] *Synonymity, generalized martingales and subfiltrations*, **Annals of Probability**, vol. 12, pp. 703–713.

[1987] *A characterization of adapted distribution*, **Annals of Probability**. vol. 15, pp. 1600–1611.

[1990] *Enlarging saturated filtrations*, preprint.

[1991] *Convergence in distribution and skorokhod convergence for the general theory of processes*, **Probability Theory and Related Fields**, vol. 89, pp. 239–259.

[1992] *Extending probability spaces and adapted distributions*, **Seminaire de probabilites xxvi** (J. Azema et al., editor), Lecture Notes in Mathematics, Springer-Verlag, pp. 560–574.

DOUGLAS HOOVER AND H. JEROME KEISLER

[1984] *Adapted probability distributions*, **Transactions of the American Mathematical Society**, vol. 286, pp. 159–201.

JEAN JACOD

[1979] **Calcul stochastique et problémes de martingales**, Lecture Notes in Mathematics, vol. 714, Springer-Verlag.

RENLING JIN AND H. JEROME KEISLER

[2000] *Maharam spectra of loeb spaces*, **The Journal of Symbolic Logic**, vol. 65, pp. 550–566.

H. JEROME KEISLER

[1977] *Hyperfinite model theory*, **Logic colloquium 76** (R. Gandy and J.M.E. Hyland, editors), North-Holland, pp. 5–110.

[1979] *Hyperfinite probability theory and probability logic*, Seminar notes. University of Wisconsin-Madison.

[1984] **An infinitesimal approach to stochastic analysis**, American Mathematical Society Memoirs, vol. 397.

[1985] *Probability quantifiers*, **Model theoretic logics** (Jon Barwise Solomon Feferman, editor), Springer-Verlag, pp. 509–556.

[1986a] *A completeness proof for adapted probability logic*, **Annals of Pure and Applied Logic**, vol. 31, pp. 61–70.

[1986b] *Hyperfinite models of adapted probability logic*, **Annals of Pure and Applied Logic**, vol. 31, pp. 71–86.

[1988] *Infinitesimals in probability theory*, **Nonstandard analysis and its applications** (Nigel Cutland, editor), LSMGT, no. 10, Cambridge University Press, pp. 106–139.

[1991] *From discrete to continuous time*, **Annals of Pure and Applied Logic**, vol. 52, pp. 99–141.

[1994] *The hyperreal line*, **Real numbers, generalizations of the reals, and theories of continua** (P. Erlich, editor), Kluwer, pp. 207–237.

[1995] *A neometric survey*, **Developments in nonstandard analysis** (Nigel Cutland et al., editor), Longman, pp. 233–250.

[1997a] *Rich and saturated adapted spaces*, **Advances in Mathematics**, vol. 128, pp. 242–288.

[1997b] *Solution of stochastic differential equations with extra properties*, **Nonstandard analysis, theory and applications** (L. Arkeryd et al., editor), NATO Advanced Studies Institute, Kluwer, pp. 259–278.

[1998] *Quantifier elimination for neocompact sets*, **The Journal of Symbolic Logic**, vol. 63, pp. 1442–1472.

H. JEROME KEISLER AND YENENG SUN

[2001] *Loeb measures and borel algebras*, **Reuniting the antipodes—constructive and nonstandard views of the continuum. proceedings of the symposion in san servolo/venice, italy, may 17-22, 1999** (H. Osswald and P. Schuster, editors), Kluwer.

FRANK KNIGHT

[1975] *A predictive view of continuous time processes*, **Annals of Probability**, vol. 3, pp. 573–596.

JOHN LAMPERTI

[1977] **Stochastic processes**, Applied Mathematical Sciences, vol. 23, Springer-Verlag.

TOM LINDSTRØM

[1988] *An invitation to nonstandard analysis*. **Nonstandard analysis an its applications** (Nigel Cutland, editor), London Mathematical Society Student Texts, no. 10, Cambridge University Press, pp. 1–105.

PETER LOEB

[1975] *Conversion from standard to nonstandard measure spaces and applications in probability theory*, **Transactions of the American Mathematical Society**, vol. 211, pp. 113–122.

PAUL LOÈVE

[1977-1978] **Probability theory**, Springer-Verlag.

MIGUEL LÓPEZ

[1989] *Distribuciones adaptadas en probabilidad. undergraduate thesis, department of mathematics*, **Master's thesis**. Universidad de Los Andes. Bogotá. Colombia.

DOROTHY MAHARAM

[1942] *On homogeneous measure algebras*, **Proceedings of the National Academy of Sciences of the U.S.A.**, vol. 28, pp. 108–111.

[1950] *Decompositions of measure algebras and spaces*, **Transactions of the American Mathematical Society**, vol. 69, pp. 143–160.

[1958] *Automorphisms of products of measure spaces*, **Proceedings of the American Mathematical Society**, vol. 9, pp. 702–707.

JAVIER PEÑA

[1993] *Games between stochastic processes and a universality theorem for two-parameter stochastic processes*, **Master's thesis**, Universidad de Los Andes. Bogotá, Colombia.

MIODRAG RAŠKOVIC AND RADOSAV DORDEVIC

[1996] *Probability quantifiers and operators*, Vesta-Belgrade.

EMILIO REMOLINA

[1993] *Some topics in model theory of stochastic processes*, **Master's thesis**, Universidad de Los Andes. Bogotá, Colombia.

KEITH STROYAN AND JOSÉ BAYOD

[1986] *Foundations of infinitesimal stochastic analysis*, Studies in Logic and the Foundations of Mathematics, vol. 119, North-Holland.

DAVID WILLIAMS

[1991] *Probability with martingales*, Cambridge Mathematical Textbooks, Cambridge University Press.

Index

LECTURE NOTES IN LOGIC
General Remarks

This series is intended to serve researchers, teachers, and students in the field of symbolic logic, broadly interpreted. The aim of the series is to bring publications to the logic community with the least possible delay and to provide rapid dissemination of the latest research. Scientific quality is the overriding criterion by which submissions are evaluated.

Books in the Lecture Notes in Logic series are printed by photo-offset from master copy prepared using LaTeX or (preferably) $\mathcal{AMS}$-LaTeX and the ASL style files. For this purpose the Association for Symbolic Logic provides technical instructions to authors. Careful preparation of manuscripts will help keep production time short, reduce costs, and ensure quality of appearance of the finished book. Authors receive 50 free copies of their book. No royalty is paid on LNL volumes.

Commitment to publish may be made by letter of intent rather than by signing a formal contract, at the discretion of the ASL Publisher. The Association for Symbolic Logic secures the copyright for each volume.

The editors prefer email contact and encourage electronic submissions.

Editorial Board

Editorial Policy

1. Submissions are invited in the following categories:

i) Research monographs   iii) Reports of meetings

ii) Lecture and seminar notes   iv) Texts which are out of print

Those considering a project which might be suitable for the series are strongly advised to contact the publisher or the series editors at an early stage.

2. Categories i) and ii). These categories will be emphasized by Lecture Notes in Logic and are normally reserved for works written by one or two authors. The goal is to report new developments quickly. informally, and in a way that will make them accessible to non-specialists. Books in these categories should include

– at least 100 pages of text;

– a table of contents:

– an informative introduction, perhaps with some historical remarks. which should be accessible to readers unfamiliar with the topic treated:

– a subject index.

In the evaluation of submissions. timeliness of the work is an important criterion. Texts should be well-rounded and reasonably self-contained. In most cases the work will contain results of others as well as those of the authors. In each case. the author(s) should provide sufficient motivation, examples. and applications. In this respect. Ph.D. theses will be suitable for this series only when they are of exceptional interest and of high expository quality.

Proposals for volumes in this category should be submitted (preferably in duplicate) to one of the series editors, and will be refereed. A provisional judgment on the acceptability of a project can be based on partial information about the work: a first draft. or a detailed outline describing the contents of each chapter. the estimated length. a bibliography, and one or two sample chapters. A final decision whether to accept will rest on an evaluation of the completed work.

3. Category iii). Reports of meetings will be considered for publication provided that they are of lasting interest. In exceptional cases. other multi-authored volumes may be considered in this category. One or more expert participant(s) will act as the scientific editor(s) of the volume. They select the papers which are suitable for inclusion and have them individually refereed as for a journal. Organizers should contact the Managing Editor of Lecture Notes in Logic in the early planning stages.

4. Category iv). This category provides an avenue whereby out-of-print books that are still in demand can be made available to a new generation of logicians.

5. Format. Works in English are preferred. After the manuscript is accepted in its final form. an electronic copy in LaTeX or (preferably) $\mathcal{A}_{\mathcal{M}}$S-LaTeX format will be appreciated and will advance considerably the publication date of the book. Authors are strongly urged to seek typesetting instructions from the Association for Symbolic Logic at an early stage of their manuscript preparation.

LECTURE NOTES IN LOGIC

From 1993 to 1999 this series was published under an agreement between the Association for Symbolic Logic and Springer-Verlag. Since 1999 the ASL is Publisher and A K Peters, Ltd. is Co-publisher. The ASL is committed to keeping all books in the series in print.

Current information may be found at http://www.aslonline.org, the ASL Web site. Editorial and submission policies and the list of Editors may also be found above.

Previously published books in the *Lecture Notes in Logic* are:

1. *Recursion theory.* J. R. Shoenfield. (1993, reprinted 2001; 84 pp.)

2. *Logic Colloquium '90; Proceedings of the Annual European Summer Meeting of the Association for Symbolic Logic, held in Helsinki, Finland, July 15–22, 1990.* Eds. J. Oikkonen and J. Väänänen. (1993, reprinted 2001; 305 pp.)

3. *Fine structure and iteration trees.* W. Mitchell and J. Steel. (1994; 130 pp.)

4. *Descriptive set theory and forcing: how to prove theorems about Borel sets the hard way.* A. W. Miller. (1995; 130 pp.)

5. *Model theory of fields.* D. Marker, M. Messmer, and A. Pillay. (1996; 154 pp.)

6. *Gödel '96; Logical foundations of mathematics, computer science and physics; Kurt Gödel's legacy. Brno, Czech Republic, August 1996, Proceedings.* Ed. P. Hajek. (1996, reprinted 2001; 322 pp.)

7. *A general algebraic semantics for sentential objects.* J. M. Font and R. Jansana. (1996; 135 pp.)

8. *The core model iterability problem.* J. Steel. (1997; 112 pp.)

9. *Bounded variable logics and counting.* M. Otto. (1997; 183 pp.)

10. *Aspects of incompleteness.* P. Lindstrom. (1997; 133 pp.)

11. *Logic Colloquium '95; Proceedings of the Annual European Summer Meeting of the Association for Symbolic Logic, held in Haifa, Israel, August 9–18, 1995.* Eds. J. A. Makowsky and E. V. Ravve. (1998; 364 pp.)

12. *Logic Colloquium '96; Proceedings of the Colloquium held in San Sebastian, Spain, July 9–15, 1996.* Eds. J. M. Larrazabal, D. Lascar, and G. Mints. (1998; 268 pp.)

13. *Logic Colloquium '98; Proceedings of the Annual European Summer Meeting of the Association for Symbolic Logic, held in Prague, Czech Republic, August 9–15, 1998.* Eds. S. R. Buss, P. Hájek, and P. Pudlák. (2000; 541 pp.)

14. *Model Theory of Stochastic Processes.* S. Fajardo and H. J. Keisler. (2002; 136 pp.)

15. *Reflections on the Foundations of Mathematics; Essays in honor of Solomon Feferman.* Eds. W. Seig, R. Sommer, and C. Talcott. (2002; 444 pp.)

Printed and bound by CPI Group (UK) Ltd, Croydon, CR0 4YY

23/10/2024

01777672-0008